Generative AI Foundations in Python

Discover key techniques and navigate modern challenges in LLMs

Carlos Rodriguez

Generative AI Foundations in Python

Group Product Manager: Niranjan Naikwadi

Publishing Product Manager: Tejashwini R

Book Project Manager: Hemangi Lotlikar

Senior Editor: Shrishti Pandey

Technical Editor: Rahul Limbachiya

Copy Editor: Safis Editing

Proofreader: Shrishti Pandey

Indexer: Manju Arasan

Production Designer: Alishon Mendonca

Senior DevRel Marketing Coordinator: Vinishka Kalra

First published: July 2024

Production reference: 1020724

Published by Packt Publishing Ltd.

Grosvenor House

11 St Paul's Square

Birmingham

B3 1RB, UK

ISBN 978-1-83546-082-5

www.packtpub.com

To the memory of my late friend, Austin Tribble, for exemplifying resilience and determination. To my wife, Jill Rodriguez, whose brilliance and intellectual curiosity have inspired me every day since the day we met.

– Carlos Rodriguez

Foreword

Large Language Models (**LLMs**) are poised to transform the way we interact with technology, offering unprecedented capabilities in understanding and generating human language. They have become essential tools in numerous applications, from chatbots and virtual assistants to content creation and translation services. For a subject that is extremely dynamic and complex, Carlos has managed to distill years of expertise into a work that is both accessible and comprehensive. This book not only demystifies the complexities of LLMs but also provides a comprehensive guide for practitioners and enthusiasts alike. So, it is with great pride and excitement that I pen this foreword for my good friend and esteemed colleague, Carlos Rodriguez, whose work on LLMs delves into the intricacies of model architecture, training methodologies, and practical implementations, all while maintaining a clarity that ensures readers, regardless of their background, can grasp the fundamental principles and potential applications of LLMs. Our journey together began only two short years ago; however, we found ourselves to be kindred spirits in the ever-evolving world of AI. From the outset, I was struck by Carlos' insatiable curiosity and unyielding dedication to the field of AI. Over numerous discussions and collaborative projects, I have witnessed firsthand the depth of his knowledge, the rigor of his research, and the passion that fuels his relentless pursuit of innovation. What sets *Generative AI Foundations in Python* apart is Carlos' unique ability to blend technical depth with practical insights. Each chapter is a testament to his meticulous approach and his commitment to bridging the gap between theoretical concepts and real-world solutions. Interweaving real-world examples, code snippets, and practical considerations ensures that seasoned professionals or newcomers to the field will find this book to be an invaluable resource. In closing, I invite you to embark on this journey with an open mind and a passion for learning. The landscape of LLMs is vast; there is no better guide than the one you hold in your hands. May this book inspire, educate, and ignite a passion for learning and discovery in every reader. Enjoy the journey.

– Samira Shaikh, PhD.

VP of Data Science, Artificial Intelligence, and Advanced Analytics, Popular Bank Associate Professor of Computer Science, UNC Charlotte

Contributors

About the author

Carlos Rodriguez is the Director of AI risk at a major financial institution, where he oversees the validation of cutting-edge AI and machine learning models, including generative AI, to ensure that they remain trustworthy, unbiased, and compliant with stringent regulatory standards. With a degree in data science, numerous professional certifications, and two decades of experience in emerging technology, Carlos is a recognized expert in natural language processing and machine learning. Throughout his career, he has fostered and led high-performing machine learning engineering and data science teams specializing in natural language processing and AI risk, respectively. Known for his human-centered approach to AI, Carlos is a passionate autodidact who continuously expands his knowledge as a data scientist, machine learning practitioner, and risk executive. His current focus lies in developing a comprehensive framework for evaluating generative AI models within a regulatory setting, aiming to set new industry standards for responsible AI adoption and deployment.

I want to express my gratitude to everyone who supported me throughout this process, with special thanks to my wife, Jill, for her unwavering support. I also want to extend a thank you to my parents, particularly my mother, a lifelong educator who has always encouraged me to find every opportunity to teach others. Finally, a special thanks to Morgan, Eric, Jeremy, Rose, and Samira, who so graciously took the time to review the manuscript at various stages.

About the reviewers

Morgan Boyce's education includes bachelor's degrees in economics and finance, a master's degree in mathematical finance, and a PhD in economics. He has worked in the financial services industry for nearly 20 years in various roles such as economic and quantitative research, model development, analytics, and model validation. Outside the financial services industry, Morgan's research focuses on the economics of technological innovation as well as public entrepreneurship. He also teaches various economics courses at university level.

Eric Rui is a distinguished technology and data leader in the financial services industry, renowned for his expertise in Data and AI. With a career dedicated to driving innovation and efficiency, Eric leverages cutting-edge technologies and data-driven insights to transform organizational processes. His strategic vision and technical acumen make him a key influencer, excelling in creating robust solutions that enhance data utilization and analytical capabilities. Additionally, Eric has deep knowledge of practical generative AI, applying advanced machine learning techniques to drive business growth and optimize decision-making.

Table of Contents

4

Applying Pretrained Generative Models: From Prototype to Production 71

Part 2: Practical Applications of Generative AI

5

Fine-Tuning Generative Models for Specific Tasks 105

6

Understanding Domain Adaptation for Large Language Models   121

7

Mastering the Fundamentals of Prompt Engineering   131

8

Addressing Ethical Considerations and
Charting a Path Toward Trustworthy Generative AI   149

Preface

Welcome to *Generative AI Foundations in Python: Discover key techniques and navigate modern challenges in LLMs.* This book offers an accessible introduction to generative AI and **large language models (LLMs)**, guiding the reader from core principles to practical applications. It aims to present a balanced approach, offering theory and hands-on examples, providing a strong foundation for those seeking to understand and leverage generative AI in their respective disciplines and fields.

Who this book is for

Written for data scientists, machine learning engineers, IT professionals, educators, and students with a basic grasp of machine learning and Python, the book meets the readers where they are, enabling them to engage fully with the content and build their foundational knowledge of generative AI concepts.

What this book covers

Chapter 1, Understanding Generative AI: An Introduction, lays the conceptual groundwork, broadening the reader's fundamental understanding of what this technology does, how it was derived, and how it can be used. It establishes how generative models differ from classical machine learning paradigms and elucidates how they discern complex relationships and idiosyncrasies in data to synthesize human-like text, audio, and video.

Chapter 2, Surveying GenAI Types and Modes: An Overview of GANs, Diffusers, and Transformers, explores the theoretical foundations and real-world applications of these techniques in greater depth. It dissects the architectural innovations and enhancements that improved training stability and output quality over time, bringing us to state-of-the-art LLMs.

Chapter 3, Tracing the Foundations of Natural Language Processing and the Impact of the Transformer, covers the evolution of **natural language processing** (**NLP**) that ultimately led to the advent of the Transformer architecture. It introduces the Transformer—its basis in deep learning, its self-attention architecture, and its rapid evolution, which has led to the generative AI phenomenon.

Chapter 4, Applying Pretrained Generative Models: From Prototype to Production, outlines the process of transitioning a generative AI prototype to a production-ready deployment. It walks through setting up a robust Python environment using Docker, GitHub, and CI/CD pipelines, then presents considerations for selecting and deploying a suitable pre-trained model for the project at hand, emphasizing computational considerations, proper evaluation, monitoring, and responsible AI practices.

Chapter 5, Fine-Tuning Generative Models for Specific Tasks, examines how **Parameter-Efficient Fine-Tuning** (**PEFT**) facilitates approachable continued training for specific tasks such as question-answering. It explores and defines a range of scalable fine-tuning techniques, comparing them with other approaches such as in-context learning.

Chapter 6, Understanding Domain Adaptation for Large Language Models, introduces domain adaptation, a unique fine-tuning approach that equips models to interpret language unique to specific industries or domains, addressing the gap in LLMs' understanding of specialized language.

Chapter 7, Mastering the Fundamentals of Prompt Engineering, explores prompting techniques to examine how to adapt a general-purpose LLM without fine-tuning. It explores various prompting strategies that leverage the model's inherent capabilities to produce targeted and contextually relevant outputs. It explores a simple approach to RAG and provides techniques to understand and measure performance.

Chapter 8, Addressing Ethical Considerations and Charting a Path Toward Trustworthy Generative AI, recognizes the increasing prominence of generative AI and explores the ethical considerations that should guide its progress. It outlines key concepts such as transparency, fairness, accountability, respect for privacy, informed consent, security, and inclusivity, which are essential for the responsible development and use of these technologies.

To get the most out of this book

Readers should have a foundational understanding of Python programming and a basic grasp of machine learning concepts. Familiarity with deep learning frameworks such as TensorFlow or PyTorch will be beneficial but not essential. The book assumes an intermediate level of Python proficiency, enabling readers to focus on the generative AI concepts and applications covered throughout the chapters.

Software/hardware covered in the book	Operating system requirements
Python 3	GPU-enabled Windows, macOS, or Linux

The book's coding examples are designed to be compatible with Python 3 and run on Windows, macOS, or Linux operating systems. To fully engage with the hands-on tutorials and examples, access to a GPU is recommended, as many generative AI models are computationally intensive. The book provides guidance on setting up a suitable development environment, including instructions for installing necessary libraries and dependencies.

If you are using the digital version of this book, we advise you to type the code yourself or access the code from the book's GitHub repository (a link is available in the next section). Doing so will help you avoid any potential errors related to the copying and pasting of code.

Throughout the book, readers are encouraged to actively experiment with the code samples provided and adapt them to their own projects. The companion GitHub repository serves as a valuable resource, offering more complete and modular versions of the code examples presented in the chapters. Accessing and working with this code will enhance the reader's learning experience and help solidify their understanding of the concepts covered.

Download the example code files

You can download the example code files for this book from GitHub at `https://github.com/PacktPublishing/Generative-AI-Foundations-in-Python`. Any code updates will be provided in the GitHub repository. Please feel free to open issues on this repository should any arise.

We also have other code bundles from our rich catalog of books and videos available at `https://github.com/PacktPublishing/`. Check them out!

Conventions used

There are a number of text conventions used throughout this book.

`Code in text`: Indicates code words in text, database table names, folder names, filenames, file extensions, pathnames, dummy URLs, user input, and Twitter handles. Here is an example: "Each entry in the dataset needs to be tokenized and structured with the necessary fields such as `input_ids` and `attention_mask`."

A block of code is set as follows:

```
# Get the start and end positions
answer_start_scores = outputs.start_logits
answer_end_scores = outputs.end_logits
```

Bold: Indicates a new term, an important word, or words that you see onscreen. For instance, words in menus or dialog boxes appear in **bold**. Here is an example: "Click the + icon in the top-right corner of the GitHub home page and select **New repository**."

> Tips or important notes
> Appear like this.

Get in touch

Feedback from our readers is always welcome.

General feedback: If you have questions about any aspect of this book, email us at `customercare@packtpub.com` and mention the book title in the subject of your message.

Errata: Although we have taken every care to ensure the accuracy of our content, mistakes do happen. If you have found a mistake in this book, we would be grateful if you would report this to us. Please visit www.packtpub.com/support/errata and fill in the form.

Piracy: If you come across any illegal copies of our works in any form on the internet, we would be grateful if you would provide us with the location address or website name. Please contact us at copyright@packt.com with a link to the material.

If you are interested in becoming an author: If there is a topic that you have expertise in and you are interested in either writing or contributing to a book, please visit authors.packtpub.com.

Share Your Thoughts

Once you've read *Generative AI Foundations in Python*, we'd love to hear your thoughts! Scan the QR code below to go straight to the Amazon review page for this book and share your feedback.

https://packt.link/r/1-835-46082-8

Your review is important to us and the tech community and will help us make sure we're delivering excellent quality content.

Download a free PDF copy of this book

Thanks for purchasing this book!

Do you like to read on the go but are unable to carry your print books everywhere?

Is your eBook purchase not compatible with the device of your choice?

Don't worry, now with every Packt book you get a DRM-free PDF version of that book at no cost.

Read anywhere, any place, on any device. Search, copy, and paste code from your favorite technical books directly into your application.

The perks don't stop there, you can get exclusive access to discounts, newsletters, and great free content in your inbox daily

Follow these simple steps to get the benefits:

1. Scan the QR code or visit the link below

https://packt.link/free-ebook/9781835460825

2. Submit your proof of purchase

3. That's it! We'll send your free PDF and other benefits to your email directly

Part 1:
Foundations of Generative AI and the Evolution of Large Language Models

This part provides an overview of generative AI and the role of large language models. It covers the basics of generative AI, different types of generative models, including GANs, diffusers, and transformers, and the foundational aspects of natural language processing. Additionally, it explores how pretrained generative models can be applied from prototype to production, setting the stage for more advanced topics.

This part contains the following chapters:

- *Chapter 1, Understanding Generative AI: An Introduction*

- *Chapter 2, Surveying GenAI Types and Modes: An Overview of GANs, Diffusers, and Transformers*

- *Chapter 3, Tracing the Foundations of Natural Language Processing and the Impact of the Transformer*

- *Chapter 4, Applying Pretrained Generative Models: From Prototype to Production*

1

Understanding Generative AI: An Introduction

In his influential book *The Singularity Is Near* (2005), renowned inventor and futurist Ray Kurzweil asserted that we were on the precipice of an exponential acceleration in technological advancements. He envisioned a future where technological innovation would continue to accelerate, eventually leading to a **singularity**—a point where **artificial intelligence** (**AI**) could transcend human intelligence, blurring the lines between humans and machines. Fast-forward to today and we find ourselves advancing along the trajectory Kurzweil outlined, with generative AI marking a significant stride along this path. Today, we are experiencing state-of-the-art generative models can behave as collaborators capable of synthetic understanding and generating sophisticated responses that mirror human intelligence.. The rapid and exponential growth of generative approaches is propelling Kurzweil's vision forward, fundamentally reshaping how we interact with technology.

In this chapter, we lay the conceptual groundwork for anyone hoping to apply generative AI to their work, research, or field of study, broadening a fundamental understanding of what this technology does, how it was derived, and how it can be used. It establishes how generative models differ from classical **machine learning** (**ML**) paradigms and elucidates how they discern complex relationships and idiosyncrasies in data to synthesize human-like text, audio, and video. We will explore critical foundational generative methods, such as generative adversarial networks (GANs), diffusion models, and transformers, with a particular emphasis on their real-world applications.

Additionally, this chapter hopes to dispel some common misunderstandings surrounding generative AI and provides guidelines to adopt this emerging technology ethically, considering its environmental footprint and advocating for responsible development and adoption. We will also highlight scenarios where generative models are apt for addressing business challenges. By the conclusion of this chapter, we will better understand the potential of generative AI and its applications across a wide array of sectors and have critically assessed the risks, limitations, and long-term considerations.

Whether your interest is casual, you are a professional transitioning from a different field, or you are an established practitioner in the fields of data science or ML, this chapter offers a contextual understanding to make informed decisions regarding the responsible adoption of generative AI.

Ultimately, we aim to establish a foundation through an introductory exploration of generative AI and **large language models (LLMs)**, dissected into two parts.

The beginning of the book will introduce the fundamentals and history of generative AI, surveying various types, such as GANs, diffusers, and transformers, tracing the foundations of **natural language generation (NLG)**, and demonstrating the basic steps to implement generative models from prototype to production. Moving forward, we will focus on slightly more advanced application fundamentals, including fine-tuning generative models, prompt engineering, and addressing ethical considerations toward the responsible adoption of generative AI. Let's get started.

Generative AI

In recent decades, AI has made incredible strides. The origins of the field stem from classical statistical models meticulously designed to help us analyze and make sense of data. As we developed more robust computational methods to process and store data, the field shifted—intersecting computer science and statistics and giving us ML. ML systems could learn complex relationships and surface latent insights from vast amounts of data, transforming our approach to statistical modeling.

This shift laid the groundwork for the rise of deep learning, a substantial step forward that introduced multi-layered neural networks (i.e., a system of interconnected functions) to model complex patterns. Deep learning enabled powerful discriminative models that became pivotal for advancements in diverse fields of research, including image recognition, voice recognition, and natural language processing.

However, the journey continues with the emergence of generative AI. Generative AI harnesses the power of deep learning to accomplish a broader objective. Instead of classifying and discriminating data, generative AI seeks to learn and replicate data distributions to "create" entirely new and seemingly original data, mirroring human-like output.

Distinguishing generative AI from other AI models

Again, the critical distinction between discriminative and generative models lies in their objectives. Discriminative models aim to predict target outputs given input data. Classification algorithms, such as logistic regression or support vector machines, find decision boundaries in data to categorize inputs as belonging to one or more class. Neural networks learn input-output mappings by optimizing weights through backpropagation (or tracing back to resolve errors) to make accurate predictions. Advanced gradient boosting models, such as XGBoost or LightGBM, further enhance these discriminative models by employing decision trees and incorporating the principles of gradient boosting (or the strategic ensembling of models) to make highly accurate predictions.

Generative methods learn complex relationships through expansive training in order to generate new data sequences enabling many downstream applications. Effectively, these models create synthetic outputs by replicating the statistical patterns and properties discovered in training data, capturing nuances and idiosyncrasies that closely reflect human behaviors.

In practice, a discriminative image classifier labels images containing a cat or a dog. In contrast, a generative model can synthesize diverse, realistic cat or dog images by learning the distributions of pixels and implicit features from existing images. Moreover, generative models can be trained across modalities to unlock new possibilities in synthesis-focused applications to generate human-like photographs, videos, music, and text.

There are several key methods that have formed the foundation for many of the recent advancements in Generative AI, each with unique approaches and strengths. In the next section, we survey generative advancements over time, including adversarial networks, variational autoencoders, diffusion models, and autoregressive transformers, to better understand their impact and influence.

Briefly surveying generative approaches

Modern generative modeling encompasses diverse architectures suited to different data types and distinct tasks. Here, we briefly introduce some of the key approaches that have emerged over the years, bringing us to the state-of-the-art models:

- **Generative adversarial networks (GANs)** involve two interconnected neural networks—one acting as a generator to create realistic synthetic data and the other acting as a discriminator that distinguishes between real and synthetic (fake) data points. The generator and discriminator are adversaries in a **zero-sum game**, each fighting to outperform the other. This adversarial relationship gradually improves the generator's capacity to produce vividly realistic synthetic data, making GANs adept at creating intricate image distributions and achieving photo-realistic image synthesis.

- **Variational autoencoders (VAEs)** employ a unique learning method to compress data into a simpler form (or latent representation). This process involves an encoder and a decoder that work conjointly (Kingma & Welling, 2013). While VAEs may not be the top choice for image quality, they are unmatched in efficiently separating and understanding complex data patterns.

- **Diffusion models** continuously add Gaussian noise to data over multiple steps to corrupt it. Gaussian noise can be thought of as random variations applied to a signal to distort it, creating "noise". Diffusion models are trained to eliminate the added noise to recover the original data distribution. This type of reverse engineering process equips diffusion models to generate diverse, high-quality samples that closely replicate the original data distribution, producing diverse high-fidelity images (Ho et al., 2020).

- **Autoregressive transformers** leverage parallelizable self-attention to model complex sequential dependencies, showing exceptional performance in language-related tasks (Vaswani et al., 2017). Pretrained models such as GPT-4 or Claude have demonstrated the capability for generalizations in natural language tasks and impressive human-like text generation. Despite ethical issues and misuse concerns, transformers have emerged as the frontrunners in language modeling and multimodal generation.

Collectively, these methodologies paved the way for advanced generative modeling across a wide array of domains, including images, videos, audio, and text. While architectural and engineering innovations progress daily, generative methods showcase unparalleled synthesis capabilities across diverse modalities. Throughout the book, we will explore and apply generative methods to simulate real-world scenarios. However, before diving in, we further distinguish generative methods from traditional ML methods by addressing some common misconceptions.

Clarifying misconceptions between discriminative and generative paradigms

To better understand the distinctive capabilities and applications of traditional ML models (often referred to as discriminative) and generative methods, here, we clear up some common misconceptions and myths:

Myth 1: Generative models cannot recognize patterns as effectively as discriminative models.

Truth: State-of-the-art generative models are well-known for their impressive abilities to recognize and trace patterns, rivaling some discriminative models. Despite primarily focusing on creative synthesis, generative models display classification capabilities. However, the classes output from a generative model can be difficult to explain as generative models are not explicitly trained to learn decision boundaries or predetermined relationships. Instead, they may only learn to simulate classification based on labels learned implicitly (or organically) during training. In short, in cases where the explanation of model outcomes is important, classification using a discriminative model may be the better choice.

Example: Consider GPT-4. In addition to synthesizing human-like text, it can understand context, capture long-range dependencies, and detect patterns in texts. GPT-4 uses these intrinsic language processing capabilities to discriminate between classes, such as traditional classifiers. However, because GPT learns semantic relationships through extensive training, explaining its decision-making cannot be accomplished using any established methods.

Myth 2: Generative AI will eventually replace discriminative AI.

Truth: This is a common misunderstanding. Discriminative models have consistently been the option for high-stakes prediction tasks because they focus directly on learning the decision boundary between classes, ensuring high precision and reliability. More importantly, discriminative models can be explained post-hoc, making them the ultimate choice for critical applications in sectors such as healthcare, finance, and security. However, generative models may increasingly become more popular for high-stakes modeling as explainability techniques emerge.

Example: Consider a discriminative model trained specifically for disease prediction in healthcare. A specialized model can classify data points (e.g., images of skin) as healthy or unhealthy, giving healthcare professionals a tool for early intervention and treatment plans. Post-hoc explanation methods, such as SHAP, can be employed to identify and analyze the key features that influence classification outcomes. This approach offers clear insights into the specific results (i.e., feature attribution).

Myth 3: Generative models continuously learn from user input.

Truth: Not exactly. Generative LLMs are trained using a static approach. This means they learn from a vast training data corpora, and their knowledge is limited to the information contained within that training window. While models can be augmented with additional data or in-context information to help them contextualize, giving the impression of real-time learning, the underlying model itself is essentially frozen and does not learn in real time.

Example: GPT-3 was trained in 2020 and only contained information up to that date until its successor GPT-3.5, released in March of 2023. Naturally, GPT-4 was trained on more recent data, but due to training limitations (including diminishing performance returns), it is reasonable to expect that subsequent training checkpoints will be released periodically and not continuously.

While generative and discriminative models have distinct strengths and limitations, knowing when to apply each paradigm requires evaluating several key factors. As we have clarified some common myths about their capabilities, let's turn our attention to guidelines for selecting the right approach for a given task or problem.

Choosing the right paradigm

The choice between generative and discriminative models depends on various factors, such as the task or problem at hand, the quality and quantity of data available, the desired output, and the level of performance required. The following is a list of key considerations:

- **Task specificity**: Discriminative models are more suitable for high-stakes applications, such as disease diagnosis, fraud detection, or credit risk assessment, where precision is crucial. However, generative models are more adept at creative tasks such as synthesizing images, text, music, or video.

- **Data availability**: Discriminative models tend to overfit (or memorize examples) when trained on small datasets, which may lead to poor generalization. On the other hand, because generative models are often pretrained on vast amounts of data, they can produce a diverse output even with minimal input, making them a viable choice when data are scarce.

- **Model performance**: Discriminative models outperform generative models in tasks where it is crucial to learn and explain a decision boundary between classes or where expected relationships in the data are well understood. Generative models usually excel in less constrained tasks that require a measure of perceived creativity and flexibility.

- **Model explainability**: While both paradigms can include models that are considered "black boxes" or not intrinsically interpretable, generative models can be more difficult, or at times, impossible to explain, as they often involve complex data generation processes that rely on understanding the underlying data distribution. Alternatively, discriminative models often focus on learning the boundary between classes. In use cases where model explainability is a key requirement, discriminative models may be more suitable. However, generative explainability research is gaining traction.

- **Model complexity**: Generally, discriminative models require less computational power because they learn to directly predict some output given a well-defined set of inputs.

  Alternatively, generative models may consume more computational resources, as their training objective is to jointly capture the intricate hidden relationships between both inputs and presumed outputs. Accurately learning these intricacies requires vast amounts of data and large computations. Computational efficiency in generative LLM training (e.g., quantization) is a vibrant area of research.

Ultimately, the choice between generative and discriminative models should be made by considering the trade-offs involved. Moreover, the adoption of these paradigms requires different levels of infrastructure, data curation, and other prerequisites. Occasionally, a hybrid approach that combines the strengths of both models can serve as an ideal solution. For example, a pretrained generative model can be fine-tuned as a classifier. We will learn about task-specific fine-tuning in *Chapter 5*.

Now that we have explored the key distinctions between traditional ML (i.e., discriminative) and generative paradigms, including their distinct risks, we can look back at how we arrived at this paradigm shift. In the next section, we take a brief look at the evolution of generative AI.

Looking back at the evolution of generative AI

The field of generative AI has experienced an unprecedented acceleration, leading to a surge in the development and adoption of foundation models such as GPT. However, this momentum has been building for several decades, driven by continuous and significant advancements in ML and natural language generation research. These developments have brought us to the current generation of state-of-the-art models.

To fully appreciate the current state of generative AI, it is important to understand its evolution, beginning with traditional language processing techniques and moving through to more recent advancements.

Overview of traditional methods in NLP

Natural language processing (**NLP**) technology has enabled machines to understand, interpret, and generate human language. It emerged from traditional statistical techniques such as n-grams and **hidden Markov models** (**HMMs**), which converted linguistic structures into mathematical models that machines could understand.

Initially, n-grams and HMMs were the primary methods used in NLP. N-grams predicted the next word in a sequence based on the last "*n*" words, while HMMs modeled sequences by considering every word as a state in a Markov process. These early methods were good at capturing local patterns and short-range dependencies in language.

As computational power and data availability grew, more sophisticated techniques for natural language processing emerged. Among these was the **recurrent neural network** (**RNN**), which managed relationships across extended sequences and was proven to be effective in tasks where prior context influenced future predictions.

Subsequently, **long short-term memory networks** (**LSTMs**) were developed.

Unlike traditional RNNs, LSTMs had a unique ability to retain relevant long-term information while disregarding irrelevant data, maintaining semantic relationships across prolonged sequences.

Further advancements led to the introduction of sequence-to-sequence models, often utilizing LSTMs as their underlying structure. These models revolutionized fields such as machine translation and text summarization by dramatically improving efficiency and effectiveness.

Overall, NLP evolved from traditional statistical methods to advanced neural networks, transforming how we interacted with machines and enabling countless applications, such as machine translation and information retrieval (IR) (or finding relevant text based on a query). As the NLP field matured, incorporating the strengths of traditional statistical methods and advanced neural networks, a renaissance was forming. The next generation of NLP advancements would introduce transformer architectures, starting with the seminal paper *Attention is All You Need* and later the release of models such as BERT and eventually GPT.

Arrival and evolution of transformer-based models

The release of the research paper titled *Attention is All You Need* in 2017 served as a paradigm shift in natural language processing. This pivotal paper introduced the transformer model, an architectural innovation that provided an unprecedented approach to sequential language tasks such as translation. The transformer model contrasted with prior models that processed sequences serially. Instead, it simultaneously processed different segments of an input sequence, determining its relevance based on the task. This innovative processing addressed the complexity of long-range dependencies in sequences, enabling the model to draw out the critical semantic information needed for a task. The transformer was such a critical advancement that nearly every state-of-the-art generative LLM applies some derivation of the original architecture. Its importance and influence motivate our detailed exploration and implementation of the original transformer in *Chapter 3*.

With the transformer came significant advancements in natural language processing, including GPT-1 or Generative Pretrained Transformer 1 (Radford et al., 2018). GPT-1 introduced a novel directional architecture to tackle diverse NLP tasks.

Coinciding with GPT-1 was **BERT**, or **bidirectional encoder representations from transformers**, a pioneering work in the family of transformer-based models. BERT stood out among its predecessors, analyzing sentences forward and backward (or bi-directionally). This bidirectional analysis allowed BERT to capture semantic and syntactic nuances more effectively. At the time, BERT achieved unprecedented results when applied to complex natural language tasks such as named entity recognition, question answering, and sentiment analysis (Devlin et al., 2018).

Later, GPT-2, the much larger successor to GPT-1, attracted immense attention, as it greatly outperformed any of its predecessors across various tasks. In fact, GPT-2 was so unprecedented in its ability to generate human-like output that concerns about potential implications led to a delay in its initial release (Hern, 2019).

Amid early concerns, OpenAI followed up with the development of GPT-3, signaling a leap in the potential of LLMs. Developers demonstrated the potential of training at a massive scale, reaching 175 billion parameters (or adjustable variables learned during training), surpassing its two predecessors. GPT-3 was a "general-purpose" learner, capable of performing a wide range of natural language tasks learned implicitly from its training corpus instead of through task-specific fine-tuning. This capability sparked the exploration of foundation model development for general use across various domains and tasks. GPT-3's distinct design and unprecedented scale led to a generation of generative models that could perform an indefinite number of increasingly complex downstream tasks learned implicitly through its extensive training.

Development and impact of GPT-4

OpenAI's development of GPT-4 marked a significant advance in the potential of large-scale, multimodal models. GPT-4, capable of processing image and text inputs and producing text outputs, represented yet another giant leap ahead of predecessors.

GPT-4 exhibited human-level performance on various professional and academic benchmarks. For instance, it passed a simulated bar exam with a score falling into the top 10% of test-takers (OpenAI, 2023).

A key distinction of GPT-4 is what happens after pretraining. Open AI applied **reinforcement learning with human feedback (RLHF)**—a type of risk/reward training derived from the same technique used to teach autonomous vehicles to make decisions based on the environment they encounter. In the case of GPT-4, the model learned to respond appropriately to a myriad of scenarios, incorporating human feedback along the way. This novel refinement strategy drastically improved the model's propensity for factuality and its adherence to desired behaviors. The integration of RLHF demonstrated how models could be better aligned with human judgment toward the goal of responsible AI.

However, despite demonstrating groundbreaking abilities, GPT-4 had similar limitations to earlier GPT models. It was not entirely reliable and had a limited context window (or input size). Meaning it could not receive large texts or documents as input. It was also prone to hallucination. As discussed, Hallucination is an anthropomorphized way of describing the model's tendency to generate content that is not grounded in fact or reality. A hallucination occurs because generative language models (without augmentation) synthesize content purely based on semantic context and don't perform any logical processing to verify factuality. This weakness presented meaningful risks, particularly in contexts where fact-based outcomes are paramount.

Despite limitations, GPT-4 made significant strides in language model performance. As with prior models, GPT-4's development and potential use underscored the importance of safety and ethical considerations for future AI applications. As a result, the rise of GPT-4 accentuated the ongoing discussions and research into the potential implications of deploying such powerful models. In the next section, we briefly survey some of the known risks that are unique to generative AI.

Looking ahead at risks and implications

Both generative and discriminative AI introduce unique risks and benefits that must be weighed carefully. However, generative methods can not only carry forward but also exacerbate many risks associated with traditional ML while also introducing new risks. Consequently, before we can adopt generative AI in the real world and at scale, it is essential to understand the risks and establish responsible governance principles to help mitigate them:

- **Hallucination**: This is a term widely used to describe when models generate factually inaccurate information. Generative models are adept at producing plausible-sounding output without basis in fact. As such, it is critical to ground generative models with factual information. The term "grounding" refers to appending model inputs with additional information that is known to be factual. We explore grounding techniques in *Chapter 7*. Additionally, it is essential to have a strategy for evaluating model outputs that includes human review.

- **Plagiarism**: Since generative models are sometimes trained on uncrated datasets, some training corpora may have included data without explicit permissions. Models may produce information that is subject to copyright protections or can be claimed as intellectual property.

- **Accidental memorization**: As with many ML models that train on immense corpora, generative models tend to memorize parts of the training data. In particular, they are prone to memorizing sparse examples that do not fit neatly into a broader pattern. In some cases, models could memorize sensitive information that can be extracted and exposed (Brundage et al., 2020; Carlini et al., 2020). Consequently, whether consuming a pretrained model or fine-tuning (i.e., continued model training), training data curation is essential.

- **Toxicity and bias**: Another byproduct of large-scale model training is that the model will inevitably learn any societal biases embedded in the training data. Biases can manifest as gender, racial, or socioeconomic biases in generated text or images, often replicating or amplifying stereotypes. We detail mitigations for this risk in *Chapter 8*.

With an understanding of some of the risks, we turn our focus to the nuanced implications of adopting generative AI:

- **Ethical**: As discussed, these models inevitably learn and reproduce the biases inherent in the training data, raising serious ethical questions. Similarly, concerns about data privacy and security have emerged due to the model's susceptibility to memorizing and exposing its training data. This has led to calls for robust ethical guidelines and data privacy regulations (Gebru et al., 2018).

- **Environmental**: LLMs are computational giants, demanding unprecedented resources for training and implementation. Thus, they inevitably present environmental impacts. The energy consumption required to train an LLM produces substantial carbon dioxide emissions—roughly the equivalent lifetime emissions of five vehicles. Consequently, multiple efforts are underway to increase model efficiency and reduce carbon footprints. For example, techniques such as reduced bit precision training (or quantization) and parameter efficient fine-tuning (discussed in *Chapter 5*) reduce overall training time, helping to shrink carbon footprints.

- **Social**: Along with environmental impacts, LLMs also have social implications. As these models become proficient at generating text, simulating intelligent conversation, and automating fundamental tasks, they present an unparalleled opportunity for job automation. Due to various complex factors, this potential for large-scale automation in the US may disproportionately affect marginalized or underrepresented communities. Thus, this amplifies prior concerns regarding labor rights and the need for additional protections to minimize harm.

- **Business and labor**: Along with broader socio-economic implications, we must examine more direct impacts on the business sector. While generative AI opens up new opportunities, changes in the labor market could bring about immense disruption if not addressed responsibly. Beyond labor impacts, AI advancements also significantly affect various business sectors. They can result in the creation of new roles, business models, and opportunities, requiring ongoing governance strategy and explorative frameworks that center on inclusivity, ethics, and responsible adoption.

Addressing these challenges will require technical and scientific improvements, data-specific regulations and laws, ethical guidelines, and human-centered AI governance strategies. These are integral to building an equitable, secure, and inclusive AI-driven future.

Having discussed the history, risks, and limitations of generative AI, we are now better equipped to explore the vast opportunities and applications of such transformative technology.

Introducing use cases of generative AI

Generative AI has already begun to disrupt various sectors. The technology is making waves across many disciplines, from enhancing language-based tasks to reshaping digital art. The following section offers examples of real-world applications of generative AI across different sectors:

- **Traditional natural language processing**: LLMs, such as Open AI's GPT series, have elevated traditional NLP and NLG. As discussed, these models have a unique ability to generate coherent, relevant, and human-like text. The potential of these models was demonstrated when GPT-3 outperformed classical and modern approaches in several language tasks, displaying an unprecedented understanding of human language. The release of GPT-4 and Claude 3 marked another milestone, raising the standard even further for state-of-the-art models.

- **Digital art creation**: The advent of "generative art" is evidence of the radical impact of generative AI in the field of digital art. For instance, artists can use AI generative models to create intricate designs, allowing them to focus on the conceptual aspect of art. It simplifies the process, reducing the need for high-level technical acumen.

- **Music creation**: In the music industry, generative AI can enhance the composition process. Several platforms offer high-quality AI-driven music creation tools that can generate long-form musical compositions combining different music styles across various eras and genres.

- **Streamlining business processes**: Several businesses have started employing generative AI to enable faster and more efficient processes. Generative AI-enabled operational efficiencies allow employees to focus on more strategic tasks. For example, fully integrated LLM email clients can organize emails and (combined with other technologies) learn to prioritize critical emails over time.

- **Entertainment**: While still largely experimental, LLMs show promising potential to disrupt creative writing and storytelling, particularly in the gaming industry. For example, procedural games could apply LLMs to enhance dynamic storytelling and create more engaging, personalized user experiences. As technology advances, we may see more mainstream adoption of LLMs in gaming, opening up new possibilities for interactive narratives.

- **Fashion**: In the fashion industry, generative models help designers innovate. By using a state-of-the-art generative AI model, designers can create and visualize new clothing styles by simply tweaking a few configurations.

- **Architecture and construction**: In the architectural world, generative-enhanced tools can help architects and urban planners optimize and generate design solutions, leading to more efficient and sustainable architectural designs.

- **Food industry**: Emerging AI-driven cooking assistants can generate unique food combinations, novel recipes, and modified recipes for highly specific dietary needs.

- **Education**: Generative AI-enhanced educational platforms offer the automatic creation of study aids that can facilitate personalized learning experiences and can automatically generate tailored content to accommodate specific and diverse learning styles.

However, we must balance the breadth of opportunities with sophisticated guardrails and the continued promotion of ethical use. As data scientists, policymakers, and industry leaders, we must continue to work towards fostering an environment conducive to responsible AI deployment. That said, as generative AI continues to evolve, it presents a future replete with novel innovations and applications.

The future of generative AI applications

The relentless advancement of generative AI presents a future filled with both possibilities and complex challenges. Imagine a future where a generative model trained on the world's leading climate change research can offer practical yet groundbreaking counteractive strategies with precise details about their application.

However, as we embrace an increasingly AI-centered future, we should not overlook the existing challenges. These involve the potential misuse of AI tools, unpredictable implications, and the profound ethical considerations underlying AI adoption. Additionally, sustainable and eco-conscious development is key, as training large-scale models can be resource-intensive

In an age of accelerated progress, collaboration across all stakeholders—from data scientists, AI enthusiasts, and policymakers to industry leaders—is essential. By being equipped with comprehensive oversight, robust guidelines, and strategic education initiatives, concerted efforts can safeguard a future where generative AI is ubiquitous.

Despite these hurdles, the transformative potential of generative AI remains unquestionable. With its capacity to reshape industries, redefine societal infrastructures, and alter our ways of living, learning, and working, generative AI serves as a reminder that we are experiencing a pivotal moment—one propelled by decades of scientific research and computational ingenuity that are coalescing to bring us forward as a society.

Summary

In this chapter, we traced the evolution of generative AI, distinguished it from traditional ML, explored its evolution, discussed its risks and implications, and, hopefully, dispelled some common misconceptions. We contemplated some of the possibilities anchored by consideration for its responsible adoption.

As we move on to the next chapter, we will examine the fundamental architectures behind generative AI, giving us a foundational understanding of the key generative methods, including GANs, diffusion models, and transformers. These ML methods form the backbone of generative AI and have been instrumental in bringing about the remarkable advancements we see today.

References

This reference section serves as a repository of sources referenced within this book; you can explore these resources to further enhance your understanding and knowledge of the subject matter:

- https://doi.org/10.1007/s11023-020-09526-7https://www.theguardian.com/technology/2019/feb/14/elon-musk-backed-ai-writes-convincing-news-fiction

- Brundage, M., Avin, S., Clark, J., Toner, H., Eckersley, P., Garfinkel, B., Dafoe, A., Scharre, P., Zeitzoff, T., Filar, B., Anderson, H., Roff, H., Allen, G. C., Steinhardt, J., Flynn, C., Ó hÉigeartaigh, S., Beard, S., Belfield, H., Farquhar, S., & Amodei, D. (2018). *The malicious use of artificial intelligence: Forecasting, prevention, and mitigation.* arXiv [cs.AI]. http://arxiv.org/abs/1802.07228.

- Carlini, N., Tramer, F., Wallace, E., Jagielski, M., Herbert-Voss, A., Lee, K., Roberts, A., Brown, T., Song, D., Erlingsson, U., Oprea, A., & Raffel, C. (2020). *Extracting training data from large language models.* arXiv [cs.CR]. http://arxiv.org/abs/2012.07805.

- Devlin, J., Chang, M.-W., Lee, K., & Toutanova, K. (2018). *BERT: Pre-training of deep bidirectional transformers for language understanding.* arXiv [cs.CL]. http://arxiv.org/abs/1810.04805.

- Hagendorff, T. (2020). Publisher correction to *The ethics of AI ethics: An evaluation of guidelines. Minds and Machines*, 30(3), 457–461. https://doi.org/10.1007/s11023-020-09526-7.

- Hern, A. (2019, February 14). *New AI fake text generator may be too dangerous to release, say creators. The Guardian.* `https://www.theguardian.com/technology/2019/feb/14/elon-musk-backed-ai-writes-convincing-news-fiction`.

- Ho, J., Jain, A., & Abbeel, P. (2020). *Denoising diffusion probabilistic models.* arXiv [cs.LG]. `http://arxiv.org/abs/2006.11239`.

- Kaplan, J., McCandlish, S., Henighan, T., Brown, T. B., Chess, B., Child, R., Gray, S., Kingma, D. P., & Welling, M. (2013). *Auto-encoding variational bayes.* arXiv [stat.ML]. `http://arxiv.org/abs/1312.6114`.

- Muhammad, T., Aftab, A. B., Ahsan, M. M., Muhu, M. M., Ibrahim, M., Khan, S. I., & Alam, M. S. (2022). *Transformer-based deep learning model for stock price prediction: A case study on Bangladesh stock market.* arXiv [q-fin.ST]. `http://arxiv.org/abs/2208.08300`.

- OpenAI. (2023). *GPT-4 technical report.* arXiv [cs.CL]. `http://arxiv.org/abs/2303.08774`.

- Radford, A., Wu, J., Child, R., Luan, D., Amodei, D., & Sutskever, I. (2018). *Language models are unsupervised multitask learners.*

- Vaswani, A., Shazeer, N., Parmar, N., Uszkoreit, J., Jones, L., Gomez, A. N., Kaiser, L., & Polosukhin, I. (2017). *Attention Is All You Need.* arXiv [cs.CL]. `http://arxiv.org/abs/1706.03762`.

Surveying GenAI Types and Modes: An Overview of GANs, Diffusers, and Transformers

In the previous chapter, we established the key distinction between generative and discriminative models. Discriminative models focus on predicting outputs by learning $p(output|input)$, or the conditional probability of some expected output given an input or set of inputs. In contrast, generative models, such as **Generative Pretrained Transformer** (**GPT**), generate text by predicting the next token (a partial word, whole word, or punctuation) using $p(next\ token|previous\ tokens)$, based on the probabilities of possible continuations given the current context. Tokens are represented as vectors containing embeddings that capture latent features and rich semantic dependencies learned through extensive training.

We briefly surveyed leading generative approaches, including **Generative Adversarial Networks** (**GANs**), **Variational Autoencoders** (**VAEs**), diffusion models, and autoregressive transformers. Each methodology possesses unique strengths suitable for different data types and tasks. For example, GANs are adept at generating high-fidelity photographic images through an adversarial process. Diffusion models take a probabilistic approach, iteratively adding and removing noise from data to learn robust generative representations. Autoregressive transformers leverage self-attention and massive scale to achieve remarkable controlled text generation.

In this chapter, we will explore the theoretical foundations and real-world applications of these techniques in greater depth. We will make direct comparisons, elucidating architectural innovations and enhancements that improve training stability and output quality over time. Through practical examples, we will see how researchers have adapted these models to produce art, music, videos, stories, and so on.

To enable an unbiased comparison, we will concentrate primarily on image synthesis tasks. GANs and diffusion models are specifically architected for image data, harnessing advances in convolutional processing and computer vision. Transformers, powered by self-attention, excel at language modeling but can also generate images. This will allow us to benchmark performance on a common task.

By the end of this chapter, we will have implemented state-of-the-art image generation models and explored how these core methods enhance and complement each other.

Understanding General Artificial Intelligence (GAI) Types – distinguishing features of GANs, diffusers, and transformers

The often-stunning human-like quality we experience from GAI can be attributed to deep-generative machine learning advances. In particular, three fundamental methods have inspired many derivative innovations – GANs, diffusion models, and transformers. Each has its distinct strengths and is particularly well-suited for specific applications.

We briefly described GANs, a groundbreaking approach that exploits the adversarial interplay between two competing neural networks – a generator and a discriminator – to generate hyper-realistic synthetic data. Over time, GANs have seen substantial advancements, achieving greater control in data generation, higher image fidelity, and enhanced training stability. For instance, NVIDIA's StyleGAN has created highly detailed and realistic human faces. The adversarial training process of GANs, where one network generates data and the other evaluates it, allows you to create highly refined and detailed synthetic images, enhancing realism with each training iteration. The synthetic images generated can be utilized in a plethora of domains. In the entertainment industry, they can be used to create realistic characters for video games or films. In research, they provide a means to augment datasets, especially in scenarios where real data is scarce or sensitive. Moreover, in computer vision, these synthetic images aid in training and fine-tuning other machine-learning models, advancing applications like facial recognition.

Diffusion models, an innovative generative modeling alternative, explicitly address some GAN limitations. As discussed briefly in *Chapter 1*, diffusion models adopt a unique approach to introducing and systematically removing noise, enabling high-quality image synthesis with less training complexity. In medical imaging, diffusion models can significantly enhance image clarity by generating high-resolution synthetic examples to train other machine-learning models. Introducing and then iteratively removing noise can help reconstruct high-fidelity images from lower-quality inputs, which is invaluable in scenarios where obtaining high-resolution medical images is challenging.

Simultaneously, generative transformers, initially designed for language modeling, have been adopted for multimodal synthesis. Today, transformers are not confined to language and have permeated into audio, images, and video applications. For instance, OpenAI's GPT-4 excels in processing and generating text, while DALL-E creates images from textual descriptions, a perfect example of the interplay between methods. When integrated, GPT-4 and DALL-E form a robust multimodal system. GPT-4 processes and understands textual instructions, while DALL-E takes the interpreted instructions to generate corresponding visual representations. A practical application of this combination could be automated digital advertisement creation. For example, given textual descriptions of a product and the desired aesthetic, GPT-4 could interpret these instructions, and DALL-E could generate visually compelling advertisements accordingly.

Deconstructing GAI methods – exploring GANs, diffusers, and transformers

Let's deconstruct these core approaches to understand their distinct characteristics and illustrate their transformative role in advancing generative machine learning. As GAI continues to move forward, it's crucial to understand how these approaches drive innovation.

A closer look at GANs

GANs, introduced by Goodfellow et al. in 2014, primarily consist of two neural networks – the **Generator** (**G**) and the **Discriminator** (**D**). G aims to create synthetic data resembling real data, while D strives to distinguish real from synthetic data.

In this setup, the following occurs:

1. G receives input from a "latent space," a high-dimensional space representing structured randomness. This structured randomness serves as a seed to generate synthetic data, transforming it into meaningful information.

2. D evaluates the generated data, attempting to differentiate between real (or reference) and synthetic data.

    In short, the process begins with G deriving random noise from the latent space to create data. This synthetic data, along with real data, is supplied to D, which then tries to discern between the two. Feedback from D informs the parameters of G to refine its data generation process. The adversarial interaction continues until an equilibrium is reached.

3. **Equilibrium** in GANs occurs when D can no longer differentiate between real and synthetic data, assigning an equal probability of 0.5 to both. Arriving at this state signals that the synthetic data produced by G is indistinguishable from real data, which is the core objective of the synthesis process.

Ultimately, the success of GANs has had meaningful implications for various sectors. In the automotive industry, GANs have been used to simulate real-world scenarios for autonomous vehicle testing. In the entertainment sector, GANs are deployed to generate digital characters and realistic environments for filmmaking and game design. In the art world, GANs can literally craft new words. Moreover, the development of GANs has continued to move forward over the years with significant improvements in quality, control, and overall performance.

Advancement of GANs

Since its inception, GAN technology has evolved significantly with several notable advancements:

- **Conditional GANs (cGANs)**: Introduced by Mirza and Osindero in 2014, conditional GANs incorporated specific conditions during data generation, enabling more controlled outputs. cGANs have been used in tasks such as image-to-image translation (e.g., converting photos into paintings).

- **Deep Convolutional GANs (DCGANs)**: In 2015, Radford et al. enhanced GANs by integrating convolutional layers, which help to analyze image data in small, overlapping regions to capture fine granularity, substantially improving the visual quality of the synthetic output. DCGANs can generate realistic images for applications such as fashion design, where the model evolves new designs from existing trends.

- **Wasserstein GANs (WGANs)**: Introduced by Arjovsky et al. in 2017, Wasserstein GANs applied the Wasserstein distance metric to GANs' objective function, facilitating a more accurate measurement of differences between real and synthetic data. Specifically, the metric helps you find the most efficient way to make the generated data distribution resemble the real data distribution. This small adjustment leads to a more stable learning process, minimizing volatility during training. WGANs have helped generate realistic medical imagery to aid in training diagnostic AI algorithms, improving a model's ability to generalize from synthetic to actual data.

Following the advent of Wasserstein GANs, the landscape experienced a surge of inventive expansions, each tailor-made to address specific challenges or open new avenues in synthetic data generation:

- **Progressively growing GANs** incrementally increase the resolution during training, starting with lower-resolution images and gradually moving to higher resolution. This approach allows the model to learn coarse-to-fine details effectively, making training more manageable and generating high-quality images (Karras et al. 2017). These high-resolution images can enhance the realism and immersion of virtual reality environments.

- **CycleGANs** facilitates image-to-image translations, bridging domain adaptation tasks (Zhu et al., 2017). For example, a CycleGAN could transform a summer scene into a winter scene without requiring example pairs (e.g., summer-winter) during training. CycleGANs have been used to simulate weather conditions in autonomous vehicle testing, evaluating system performance under varying environmental conditions.

- **BigGANs** push the boundaries in high-resolution image generation, showcasing the versatility of GANs in complex generation tasks. They achieve this by scaling up the size of the model (more layers and units per layer) and the batch size during training, alongside other architectural and training innovations (Brock et al., 2018). BigGANs have been used to generate realistic textures for video games, enhancing gaming environments' realism.

These developments significantly broadened what GANs could achieve, ranging from high-resolution image synthesis to domain adaptation and cross-modal generation tasks. However, despite these incredible advancements, GANs have suffered from some continual limitations, which inspired alternative approaches such as diffusion.

Limitations and challenges of GANs

The training process of GANs requires a careful balance between the G and D networks. It requires substantial computational resources, often demanding powerful GPUs and enormous datasets to achieve desirable outcomes. Moreover, there are complexities in training GANs that arise from challenges such as vanishing gradients and mode collapse. While the vanishing gradient problem is a problem broadly affecting deep neural networks, mode collapse is a challenge that is particularly unique to the training of GANs. Let's explore these a bit further:

- **Vanishing gradients**: This issue arises during the neural network training phase when the gradient of the loss function diminishes to a point where the learning either drastically slows or halts. The crux of GANs lies in the delicate balance of learning between the G and D models. Disproportionate learning can hinder the overall training process. In practical terms, the issue of vanishing gradients can lead to longer training times and increased computational costs, which might render GANs impractical for time-sensitive or resource-constrained applications.

- **Mode collapse**: Inherent to GANs, mode collapse occurs when the G starts producing a narrow variety of samples, thereby stifling output diversity and undermining a network's effectiveness. Techniques such as a gradient penalty and spectral normalization have alleviated these issues. This phenomenon can significantly degrade the quality of generated data, limiting the use of GANs in applications that require diverse outputs, such as data augmentation for machine learning or generating diverse design alternatives in creative industries.

Of course, GANs carry the same ethical considerations as any state-of-the-art generative synthesis. For instance, they can be used to create deepfakes or generate biased outputs that reinforce societal prejudices. For example, when GANs, often used to generate synthetic data (e.g., faces), underrepresent certain groups, downstream applications may exhibit gender or racial bias (Kenfack et al., 2021).

Even with the advent of other generative models such as diffusion models and Transformer-based image generators, GANs have played a seminal role in shaping the trajectory of generative image synthesis, showcasing both the potential and some of the challenges inherent in this domain.

Now that we better understand GANs in the context of deep generative models, let's shift our focus to a successor in image generation, the diffusion model.

A closer look at diffusion models

Having explored the dynamics of GANs, let's transition our attention to a subsequent innovation in image generation – the diffusion model. Initially proposed by Sohl-Dickstein et al. in 2015, diffusion models present a novel approach, where a neural network iteratively introduces and subsequently removes noise from data to generate highly refined images. Unlike GANs, which leverage an adversarial mechanism involving two contrasting models, diffusion models apply a more gradual, iterative process of noise manipulation within the data.

In practical terms, GANs have shown substantial merit in art and design, creating realistic faces or generating sharp, high-fidelity images from descriptions. They are also used in data augmentation, expanding datasets by generating realistic synthetic data to augment the training of machine learning models.

Conversely, diffusion models excel in tasks requiring a structured approach to image generation, such as in medical imaging. Their iterative process can enhance the quality of medical images, such as MRI or CT scans, where noise reduction and clarity are paramount. This makes diffusion models invaluable in clinical settings, aiding in better diagnostics and analysis. Moreover, their controlled and gradual process offers a more predictable or stable training process compared to the adversarial and dynamic training of GANs.

The foundation of diffusion models is anchored in two primary processes:

- **A forward diffusion process**: This process begins with clean data (x_0) and iteratively introduces Gaussian noise, akin to progressively applying a fog-like filter, transforming the data into indistinguishable noise (x_t).

- **A learned reverse model**: Following the forward diffusion, the "reverse model" ($p\theta$) attempts to eliminate (or de-fog) the noise from the noisy data (x_t), aiming to revert to the original clean state (x_{t-1}). Specifically, this reversion is orchestrated by estimating the probability of transitioning from the noisy state back to the clear state, using a conditional distribution denoted as $p\theta(x_{t-1}|x_t)$. A **conditional distribution** tells us the likelihood of one event happening when we know another related event has occurred. In this case, the reversion estimates the likelihood of reverting to the original state, given some amount of noise.

In the pivotal work *Score-Based Generative Modeling through Stochastic Differential Equations*, the authors propose a novel framework that unifies score-based generative models and diffusion probabilistic modeling by employing **Stochastic Differential Equations** (**SDEs**). This framework involves the transformation of data distributions to a known prior distribution through the gradual addition and then removal of noise, guided by SDEs. Optimizing the reverse-time SDE – dependent only on the score of the perturbed data distribution – allows you to generate new samples. **Stochastic Gradient Descent** (**SGD**) is then applied to fine-tune the model parameters until arriving at an improved $p\theta$.

The reverse model ($p\theta$) was implemented using convolutional networks to predict variations in the Gaussian noise distribution – a critical component of the noise-introduction process within the forward diffusion. Initially, the efficacy of this approach was validated on more straightforward datasets. However, the methodology's applicability was later significantly improved to handle more complex images (Ho et al., 2020). This expansion demonstrated the practical potential of diffusion models in generating highly refined images across a broader spectrum of complexities.

Advancement of diffusion models

Since its inception, diffusion model technology has witnessed key advancements, propelling its capabilities in image generation:

- **Simplified training objectives**: Ho et al. proposed simplified training objectives that predict Gaussian noise directly, eliminating the need for conditional means and facilitating the application to more complex datasets (Ho et al., 2020). This advancement facilitated handling more complex datasets, potentially aiding in tasks such as anomaly detection or complex data synthesis, which could be resource-intensive with traditional models.

- **UNet modules with self-attention**: Ho et al. also incorporated UNet modules with self-attention into the diffusion model architecture, inspired by PixelCNN++ by Salimans et al. (2017), enhancing a model's performance on complex datasets (Ho et al., 2020). Again, enhancing performance on complex datasets facilitates better image restoration, which is particularly beneficial in fields such as medical imaging or satellite imagery analysis, where high-fidelity image reconstruction is crucial.

- **Synchronization with SDEs**: Song et al. defined diffusion models as solutions to SDEs, linking score learning with denoising score-matching losses and expanding model usage for image generation, editing, in-painting, and colorization (Song et al., 2020).

Following these foundational advancements, diffusion models witnessed a wave of innovative enhancements as researchers introduced novel methodologies to address existing challenges and broaden a model's applicability in generative modeling tasks. These advancements include the following:

- **Noise conditioning and annealing strategies**: Song et al. improved score-based models by including noise conditioning and annealing strategies, achieving performance comparable to GANs on benchmark datasets like the Flickr-Faces-HQ dataset (Song et al., 2021), which is a high-quality image dataset of human faces designed to measure GAN performance. Achieving performance comparable to GANs could make diffusion models a viable alternative for high-fidelity image generation tasks in areas where GANs are traditionally used.

- **Latent Diffusion Models** (**LDMs**): Rombach et al. addressed computational inefficiency by proposing LDMs, which operate in a compressed latent space learned by autoencoders, employing perceptual losses to create a visually equivalent, reduced latent space (Rombach et al., 2021). By addressing computational inefficiency, LDMs could expedite the image generation process, making them suitable for real-time applications or scenarios where computational resources are limited.

- **Classifier-free guidance**: Ho & Salimans introduced classifier-free guidance for controlled generation without relying on pre-trained networks, marking a step toward more flexible generation techniques (Ho & Salimans, 2022). This advancement led to more flexible generation techniques, enabling more controlled and customized image generation in applications such as design, advertising, or content creation without relying on pre-trained networks.

Subsequent explorations in the diffusion model domain extended its applications, showcasing versatility:

- **Video generation**: Ho et al. adapted diffusion models for video generation, demonstrating their utility beyond static image generation (Ho et al., 2022)

- **3D data processing**: Luo & Hu extended the application to 3D data processing, showcasing the flexibility of diffusion models (Luo & Hu, 2021)

The evolution of diffusion models has led to enhanced image generation and expanded applications in video, 3D data processing, and rapid learning methodologies. However, the methodology does have its challenges and limitations, outlined in some detail in the section that follows..

Limitations and challenges of diffusion models

Despite their evident benefits and notable progress, diffusion models have some unique limitations, such as the following:

- **Sampling speed**: A notable limitation of diffusion models is the slow sampling process, particularly when compared to GANs. Sampling, in this context, refers to the process of generating new data points from the learned distribution of a model. The speed at which new samples can be generated is crucial for many real-time or near-real-time applications, and the slower sampling speed of diffusion models can be a significant drawback.

- **Stability during large-scale training**: The stability of diffusion models during large-scale training is another area requiring further exploration. Large-scale training refers to training a model on a substantial amount of data, sometimes leading to instability in the model's learning process. Ensuring stability during this phase is crucial to achieve reliable and consistent performance from the model.

A close examination of the societal impact of the media generated by these models is crucial, especially given the level of fine control now possible over the generated content. However, diffusion models' inherent simplicity, versatility, and positive inductive biases signify a bright future. These attributes suggest a trajectory of rapid development within generative modeling, potentially integrating diffusion models as pivotal components in various disciplines, such as computer vision and graphics.

A closer look at generative transformers

The revolutionary advent of transformer models has significantly impacted the task of generating high-fidelity images from text descriptions. Notable models such as **CLIP** (**Contrastive Language-Image Pretraining**) and DALL-E utilized transformers in unique ways to create images based on natural language captions. This section will discuss the transformer-based approach for text-to-image generation, its foundations, the key techniques, the resulting benefits, and some challenges.

A brief overview of transformer architecture

The original transformer architecture, introduced by Vaswani et al. in 2017, is a cornerstone of many modern language-processing systems. In fact, the transformer may be considered the most important architecture in the area of GAI, as it is foundational to the GPT series of models and many other state-of-the-art generative methods. As such, we'll cover the architecture briefly in our survey of generative approaches but will have a dedicated chapter, where we will have the opportunity to deconstruct and implement the transformer from scratch.

At the core of the transformer architecture lies the **self-attention mechanism**, a unique approach that captures complex relationships among different elements within an ordered data sequence. These elements, known as **tokens**, represent words in a sentence or characters in a word based on the level of granularity chosen for **tokenization**.

The principle of **attention** in this architecture enables a model to focus on certain pivotal aspects of the input data while potentially disregarding less significant parts. This mechanism augments the model's understanding of the context and the relative importance of words in a sentence.

The transformer bifurcates into two main segments, the **encoder** and the **decoder**, each comprising multiple layers of self-attention mechanisms. While the encoder discerns relationships between different positions in the input sequence, the decoder focuses on the outputs from the encoder, employing a variant of self-attention termed **masked self-attention** to prevent consideration of future outputs it hasn't generated yet.

The calculation of **attention weights** through the scaled dot-product of query and key vectors plays a crucial role in determining the level of focus on different parts of the input. Additionally, **multi-head attention** allows the model to channel attention toward multiple data points simultaneously.

Lastly, to retain the sequence order of data, the model adopts a strategy known as **positional encoding**. This mechanism is vital for tasks requiring an understanding of sequence or temporal dynamics, ensuring the model preserves the initial order of data throughout its processing.

Again, we will revisit the transformer architecture in *Chapter 3* to further reinforce our understanding, as it is foundational to the continued research and evolution of generative AI. Nevertheless, with at least a fundamental grasp of the Transformer architecture, we are better positioned to dissect transformer-driven generative modeling paradigms across a spectrum of applications.

Generative modeling paradigms with transformers

In tackling various tasks, transformers adopt distinct training paradigms aligning with the task at hand. For example, discriminative tasks such as classification might use a masking paradigm:

- **Masked Language Modeling** (**MLM**): MLM is a discriminative pretraining technique used by models such as **BERT** (**Bidirectional Encoder Representations from Transformers**). During training, some percentage of input tokens are randomly masked out. The model must then predict the original masked words based on the context of the surrounding unmasked words. This teaches the model to build robust context-based representations, facilitating many downstream **natural language processing** (**NLP**) tasks.

MLM, as utilized in BERT, has been instrumental in enhancing the performance of NLP systems across various domains. For instance, it can power medical coding systems in healthcare by accurately identifying and categorizing medical terms within clinical notes. This automatic coding can save significant time and reduce errors in medical documentation, thereby improving the efficiency and accuracy of healthcare data management.

For generative tasks, the focus shifts to creating new data sequences, requiring different training paradigms:

- **Sequence-to-sequence modeling**: Sequence-to-sequence models employ both an encoder and a decoder. The encoder maps the input sequence to a latent representation. The decoder then generates the target sequence token by token from that representation. This paradigm is useful for tasks such as translation, summarization, and question-answering.

- **Autoregressive modeling**: Autoregressive modeling generates sequences by predicting the next token conditioned only on previous tokens. The model produces outputs one step at a time, with each new token depending on those preceding it. Autoregressive transformers such as GPT leverage this technique for controlled text generation.

Transformers combine self-attention for long-range dependencies, pre-trained representations, and autoregressive decoding to adapt to discriminative and generative tasks.

Advanced generative synthesis can be achieved with different architectures that make trade-offs between complexity, scalability, and specialization. For example, instead of using both the encoder and decoder, many state-of-the-art generative models employ a decoder-only or encoder-only approach. The encoder-decoder framework is often the most computationally intensive learning to specialize in, as it increases model size. Decoder-only architectures leverage powerful pre-trained language models such as GPT as the decoder, reducing parameters through weight sharing. Encoder-only methods forego decoding, instead, they encode inputs and perform regression or search on the resulting embeddings. Each approach has advantages that suit certain use cases, datasets, and computational budgets. In the following sections, we explore examples of models that employ these derivative transformer architectures for creative applications, such as image generation and captioning.

Encoder-only approach

In certain models, only the encoder network maps the input to an embedding space. The output is then generated directly from this embedding, eliminating the need for a decoder. While this straightforward architecture has typically found its place in classification or regression tasks, recent advancements have broadened its application to more complex tasks. In particular, models developed for tasks such as image synthesis leverage the encoder-only setup to process both text and visual inputs, creating a multimodal relationship that facilitates the generation of high-fidelity images from natural language instruction.

Decoder-only approach

Similarly, some models operate using a decoder-only strategy, where a singular decoder network is tasked with both encoding the input and generating output. This mechanism starts by joining the input and output sequences, which the decoder processes. Despite its simplicity and the characteristic sharing of parameters between input and output stages, the effectiveness of this architecture relies heavily on the pretraining of robust decoders. Recently, even more complex tasks such as text-to-image synthesis have seen the successful deployment of the decoder-only architecture, illustrating its versatility and adaptability to diverse applications.

Advancement of transformers

Transformer mechanisms with other novel techniques to tackle generative tasks. This evolution led to distinct approaches to handling text and image generation. In this section, we will explore some of these innovative models and their unique methodologies in advancing GAI.

Encoder-decoder image generation with DALL-E

Introduced by Ramesh et al. in 2021, DALL-E employs an encoder-decoder framework to facilitate text-to-image generation. This model comprises two primary components:

- **Text encoder**: Applies the transformer's encoder, processing plain text to derive a semantic embedding that serves as the context for the image decoder.

- **Image decoder**: Applies the transformer's decoder to generate the image autoregressively, predicting each pixel based on the text embedding and previously predicted pixels.

By training on image-caption datasets, DALL-E refines the transition from text to detailed image renderings. This setup underscores the capability of dedicated encoder and decoder modules for conditional image generation.

Encoder-only image captioning with CLIP

CLIP, conceptualized by Radford et al. in 2021, adopts an encoder-only approach for image-text tasks. Key components include a visual encoder and a text encoder.

Visual Encoder and Text Encoder process the image and candidate captions, respectively, determining the matching caption based on encoded representations.

Pretraining on extensive image-text datasets enables CLIP to establish a shared embedding space, facilitating efficient inference for retrieval-based captioning.

Improving image fidelity with scaled transformers (DALL-E 2)

Ramesh et al. in 2022 extended DALL-E to DALL-E 2, showcasing techniques to enhance visual quality:

- **A scaled-up decoder**: By expanding the decoder to 3.5 billion parameters and applying classifier-free guidance during sampling, visual quality in complex image distributions such as human faces is significantly improved.

- **Hierarchical decoding for high-resolution images (GLIDE)**: Proposed by Nichol et al. in 2021, GLIDE employs a hierarchical generation strategy.

- **A coarse-to-fine approach**: This entails an initial low-resolution image prediction followed by progressive detailing through up-sampling and refining, capturing global structure and high-frequency textures.

Multimodal image generation with GPT-4

GPT-4 developed by OpenAI, is a powerful multimodal model based on the Transformer architecture. GPT-4 demonstrates a capability for conditional image generation without requiring continued training or fine-tuning:

- **Pretraining and fine-tuning**: The massive scale of GPT-4 and its pretraining on diverse datasets enable a robust understanding of relationships between textual and visual data.

- **Multimodal generation:** GPT-4 can generate images based on text descriptions. The model uses a deep neural network to encode the semantic meaning of the text into a visual representation. Given a text prompt, GPT-4 generates an image by predicting the visual content consistent with the provided text. This involves taking high-dimensional text embeddings and processing them through successive neural network layers to generate a corresponding visual representation.

Using a pretrained multimodal model eliminates the need for a separate encoder module for image inputs, facilitating rapid adaptation for image generation tasks. This approach underscores the versatility and power of Transformer architectures in generative tasks, providing a streamlined methodology to translate text into high-quality images.

Transformer architectures offer many benefits for controlled image generation when compared to GANs. Their autoregressive nature ensures precise control over image construction while allowing you to adapt to varying computational needs and diverse downstream applications. However, transformers also introduce new challenges in this domain.

Limitations and challenges of transformer-based approaches

Some early transformers-based approaches demonstrated slower sampling speed and restricted fidelity compared to GANs. Generating or manipulating images while maintaining precise control over specific attributes or characteristics of the objects within those images remains challenging. Additionally, training large-scale transformers that can overcome these challenges demands extensive computing resources. Notwithstanding, current multimodal results demonstrate a rapidly evolving and promising landscape.

We must also remember that alongside technical challenges there are broader sociotechnical implications and considerations.

Bias and ethics in generative models

Significant advancements in generative models such as GANs, diffusers, and transformers necessitate serious contemplation of potential bias and ethical implications.

We need to remain alert to the risk of reinforcing prejudices and stereotypes that reflect skewed training data. For instance, diffusion models trained on data that over-represents specific demographics might propagate these biases in their output. Analogously, language models exposed to toxic or violent content during training might generate similar content.

The directive nature of prompt-based generation also, unfortunately, opens doors to misuse if deployed carelessly. Transformers risk facilitating impersonation, misinformation, and the creation of deceptive content. Image synthesis models such as GANs could potentially be exploited to generate non-consensual deepfakes or artificial media.

Additionally, the potential for ultra-realistic output prompts ethical dilemmas regarding consent, privacy, identity, and copyright. The ability to create convincingly real yet fictional faces or voices complicates the distinction between real and synthetic, necessitating careful examination of training data sources and generative capabilities.

Further, as these technologies become ubiquitous, their societal impact must be considered. Defining clear policies will be crucial as the distinction between authentic and AI-generated content becomes increasingly ambiguous. Upholding principles of integrity, attribution, and consent remains vital.

Despite these risks, the potential benefits of generative models are substantial. Addressing bias proactively, advocating transparency, auditing data and models, and implementing safeguards become increasingly critical as technologies evolve. Ultimately, the responsibility to ensure fairness, accountability, and ethical practice falls on all developers and practitioners.

Applying GAI models – image generation using GANs, diffusers, and transformers

In this hands-on section, we'll reinforce the concepts discussed throughout the chapter by putting them into practice. You'll get a first-hand experience and deep dive into the actual implementation of generative models, specifically GANs, diffusion models, and transformers.

The Python code provided will guide you through this process. Manipulating and observing the code in action will build your understanding of the intricate workings and potential applications of these models. This exercise will provide insight into model capabilities for tasks like generating art from prompts and synthesizing hyper-realistic images.

We'll be utilizing the highly versatile `PyTorch` library, a popular choice among machine learning practitioners, to facilitate our operations. `PyTorch` provides a powerful and dynamic toolset to define and compute gradients, which is central to training these models.

In addition, we'll also use the `diffusers` library. It's a specialized library that provides functionality to implement diffusion models. This library enables us to reproduce state-of-the-art diffusion models directly from our workspace. It underpins the creation, training, and usage of denoising diffusion probabilistic models at an unprecedented level of simplicity, without compromising the models' complexity.

Through this practical session, we'll explore how to operate and integrate these libraries and implement and manipulate GANs, diffusers, and transformers using the Python programming language. This hands-on experience will complement the theoretical knowledge we have gained in the chapter, enabling us to see these models in action in the real world.

By the end of this section, you will not only have a conceptual understanding of these generative models but also understand how they are implemented, trained, and used for several innovative applications in data science and machine learning. You'll have a much deeper understanding of how these models work and the experience of implementing them yourself.

Working with Jupyter Notebook and Google Colab

Jupyter notebooks enable live code execution, visualization, and explanatory text, suitable for prototyping and data analysis. Google Colab, conversely, is a cloud-based version of Jupyter Notebook, designed for machine learning prototyping. It provides free GPU resources and integrates with Google Drive for file storage and sharing. We'll leverage Colab as our prototyping environment going forward.

Stable diffusion transformer

We begin with a pre-trained stable diffusion model, a text-to-image latent diffusion model created by researchers and engineers from CompVis, Stability AI, and LAION (Patil et al., 2022). The diffusion process is used to draw samples from complex, high-dimensional distributions, and when it interacts with the text embeddings, it creates a powerful conditional image synthesis model.

The term "stable" in this context refers to the fact that during training, a model maintains certain properties that stabilize the learning process. Stable diffusion models offer rich potential to create entirely new samples from a given data distribution, based on text prompts.

Again, for our practical example, we will Google Colab to alleviate a lot of initial setups. Colab also provides all of the computational resources needed to begin experimenting right away. We start by installing some libraries, and with three simple functions, we will build out a minimal `StableDiffusionPipeline` using a well-established open-source implementation of the stable diffusion method.

First, let's navigate to our pre-configured Python environment, Google Colab, and install the `diffusers` open-source library, which will provide most of the key underlying components we need for our experiment.

In the first cell, we install all dependencies using the following `bash` command. Note the exclamation point at the beginning of the line, which tells our environment to reach down to its underlying process and install the packages we need:

```
!pip install pytorch-fid torch diffusers clip transformers accelerate
```

Next, we import the libraries we've just installed to make them available to our Python program:

```
from typing import List
import torch
import matplotlib.pyplot as plt
from diffusers import StableDiffusionPipeline, DDPMScheduler
```

Now, we're ready for our three functions, which will execute the three tasks – loading the pre-trained model, generating the images based on prompting, and rendering the images:

```
def load_model(model_id: str) -> StableDiffusionPipeline:
    """Load model with provided model_id."""
    return StableDiffusionPipeline.from_pretrained(
        model_id,
        torch_dtype=torch.float16,
        revision="fp16",
        use_auth_token=False
    ).to("cuda")

def generate_images(
    pipe: StableDiffusionPipeline,
    prompts: List[str]
) -> torch.Tensor:
    """Generate images based on provided prompts."""
    with torch.autocast("cuda"):
        images = pipe(prompts).images
    return images

def render_images(images: torch.Tensor):
    """Plot the generated images."""
    plt.figure(figsize=(10, 5))
    for i, img in enumerate(images):
        plt.subplot(1, 2, i + 1)
        plt.imshow(img)
        plt.axis("off")
    plt.show()
```

In summary, `load_model` loads a machine learning model identified by `model_id` onto a GPU for faster processing. The `generate_images` function takes this model and a list of prompts to create our images. Within this function, you will notice `torch.autocast("cuda")`, which is a special command that allows PyTorch (our underlying machine learning library) to perform operations faster while maintaining accuracy. Lastly, the `render_images` function displays these images in a simple grid format, making use of the `matplotlib` visualization library to render our output.

With our functions defined, we select our model version, define our pipeline, and execute our image generation process:

```
# Execution
model_id = "CompVis/stable-diffusion-v1-4"
prompts = [
    "A hyper-realistic photo of a friendly lion",
    "A stylized oil painting of a NYC Brownstone"
]

pipe = load_model(model_id)
images = generate_images(pipe, prompts)
render_images(images)
```

The output in *Figure 2.1* is a vivid example of the imaginativeness and creativity we typically expect from human art, generated entirely by the diffusion process. Except, how do we measure whether the model was faithful to the text provided?

100%   50/50 [00:05<00:00, 10.20it/s]

Figure 2.1: Output for the prompts "A hyper-realistic photo of a friendly lion"
(left) and "A stylized oil painting of a NYC Brownstone" (right)

The next step is to evaluate the quality and relevance of our generated images in relation to the prompts. This is where CLIP comes into play. CLIP is designed to measure the alignment between text and images by analyzing their semantic similarities, giving us a true quantitative measure of the fidelity of our synthetic images to the prompts.

Scoring with the CLIP model

CLIP is trained to understand the relationship between text and images by learning to place similar images and text near each other in a shared space. When evaluating a generated image, CLIP checks how closely the image aligns with the textual description provided. A higher score indicates a better match, meaning the image accurately represents the text. Conversely, a lower score suggests a deviation from the text, indicating a lesser quality or fidelity to the prompt, providing a quantitative measure of how well the generated image adheres to the intended description.

Again, we will import the necessary libraries:

```
from typing import List, Tuple
from PIL import Image
import requests
from transformers import CLIPProcessor, CLIPModel
import torch
```

We begin by loading the CLIP model, processor, and necessary parameters:

```
# Constants
CLIP_REPO = "openai/clip-vit-base-patch32"

def load_model_and_processor(
    model_name: str
) -> Tuple[CLIPModel, CLIPProcessor]:
    """
    Loads the CLIP model and processor.
    """
    model = CLIPModel.from_pretrained(model_name)
    processor = CLIPProcessor.from_pretrained(model_name)
    return model, processor
```

Next, we define a processing function to adjust the textual prompts and images, ensuring that they are in the correct format for CLIP inference:

```
def process_inputs(
    processor: CLIPProcessor, prompts: List[str],
    images: List[Image.Image]) -> dict:
    """
    Processes the inputs using the CLIP processor.
```

```
"""
    return processor(text=prompts, images=images,
        return_tensors="pt", padding=True)
```

In this step, we initiate the evaluation process by inputting the images and textual prompts into the CLIP model. This is done in parallel across multiple devices to optimize performance. The model then computes similarity scores, known as logits, for each image-text pair. These scores indicate how well each image corresponds to the text prompts. To interpret these scores more intuitively, we convert them into probabilities, which indicate the likelihood that an image aligns with any of the given prompts:

```
def get_probabilities(
    model: CLIPModel, inputs: dict) -> torch.Tensor:
"""
Computes the probabilities using the CLIP model.
"""
    outputs = model(**inputs)
    logits = outputs.logits_per_image
    # Define temperature - higher temperature will make the
distribution more uniform.
    T = 10
    # Apply temperature to the logits
    temp_adjusted_logits = logits / T
    probs = torch.nn.functional.softmax(
        temp_adjusted_logits, dim=1)
    return probs
```

Lastly, we display the images along with their scores, visually representing how well each image adheres to the provided prompts:

```
def display_images_with_scores(
    images: List[Image.Image], scores: torch.Tensor) -> None:
"""
Displays the images alongside their scores.
"""
    # Set print options for readability
    torch.set_printoptions(precision=2, sci_mode=False)

    for i, image in enumerate(images):
        print(f"Image {i + 1}:")
        display(image)
        print(f"Scores: {scores[i, :]}")
        print()
```

With everything detailed, let's execute the pipeline as follows:

```
# Load CLIP model
model, processor = load_model_and_processor(CLIP_REPO)
# Process image and text inputs together
inputs = process_inputs(processor, prompts, images)
# Extract the probabilities
probs = get_probabilities(model, inputs)
# Display each image with corresponding scores
display_images_with_scores(images, probs)
```

We now have scores for each of our synthetic images that quantify the fidelity of the synthetic image to the text provided, based on the CLIP model, which interprets both image and text data as one combined mathematical representation (or geometric space) and can measure their similarity.

Scores: tensor([0.83, 0.17], grad_fn=<SliceBackward0>)

Figure 2.2: CLIP scores

For our "friendly lion," we computed scores of 83% and 17% for each prompt, which we can interpret as an 83% likelihood that the image aligns with the first prompt.

In practical scenarios, this metric can be applied across various domains:

- **Content moderation**: Automatically moderating or flagging inappropriate content by comparing images to a set of predefined descriptive prompts

- **Image retrieval**: Facilitating refined image searches by matching textual queries to a vast database of images, hence narrowing down the search to the most relevant visuals

- **Image captioning**: Assisting in generating accurate captions for images by identifying the most relevant descriptive prompts

- **Advertising**: Tailoring advertisements based on the content of images on a web page to enhance user engagement

- **Accessibility**: Enhancing accessibility features by providing accurate descriptions of images for individuals with visual impairments

This evaluation method not only speeds up processes that would otherwise require manual inspection but also lends itself to many applications that could benefit from a deeper understanding and contextual analysis of visual data. We will revisit the CLIP evaluation in *Chapter 4*, where we simulate a real-world scenario to determine the quality and appropriateness of automatically generated captions for a set of product images.

Summary

This chapter explored the theoretical underpinnings and real-world applications of leading GAI techniques, including GANs, diffusion models, and transformers. We examined their unique strengths, including GANs' ability to synthesize highly realistic images, diffusion models' elegant image generation process, and transformers' exceptional language generation capabilities.

Using a cloud-based Python environment, we implemented these models to generate compelling images and evaluated their output quality using CLIP. We analyzed how techniques such as progressive growing and classifier guidance enhanced output fidelity over time. We also considered societal impacts, urging developers to address potential harm through transparency and ethical practices.

Generative methods have unlocked remarkable creative potential, but thoughtful oversight is critical as capabilities advance. We can guide these technologies toward broadly beneficial outcomes by grounding ourselves in core methodologies, scrutinizing their limitations, and considering downstream uses. The path ahead will require continued research and ethical reflection to unlock AI's creative promise while mitigating risks.

References

This reference section serves as a repository of sources referenced within this book; you can explore these resources to further enhance your understanding and knowledge of the subject matter:

- Kenfack, P. J., Arapov, D. D., Hussain, R., Ahsan Kazmi, S. M., & Khan, A. (2021). *On the fairness of generative adversarial networks (GANs)*. `Arxiv.org`.

- Goodfellow, I., Pouget-Abadie, J., Mirza, M., Xu, B., Warde-Farley, D., Ozair, S., Courville, A., & Bengio, Y. (2014). *Generative adversarial nets. Advances in neural information processing systems*, 27.

- Nichol, A., Dhariwal, P., Ramesh, A., Shyam, P., Mishkin, P., McGrew, B., Sutskever, I., & Chen, M. (2021). *GLIDE: Towards photorealistic image generation and editing with text-guided diffusion models*. arXiv preprint arXiv:2112.10741.

- Radford, A., Kim, J. W., Hallacy, C., Ramesh, A., Goh, G., Agarwal, S., Sastry, G., Askell, A., Mishkin, P., Clark, J., Krueger, G., & Sutskever, I. (2021). *Learning Transferable Visual Models From Natural Language Supervision*. ArXiv. /abs/2103.00020.

- Ramesh, A., Pavlov, M., Goh, G., Gray, S., Voss, C., Radford, A., Chen, M., & Sutskever, I. (2022). *Hierarchical text-conditional image generation with clip latents*. arXiv preprint arXiv:2204.06125.

- Ramesh, A., Pavlov, M., Goh, G., Gray, S., Voss, C., Radford, A., Chen, M., & Sutskever, I. (2021). *Zero-shot text-to-image generation. In International Conference on Machine Learning* (pp. 8821–8831). PMLR.

- Vaswani, A., Shazeer, N., Parmar, N., Uszkoreit, J., Jones, L., Gomez, A. N., Kaiser, L., & Polosukhin, I. (2017). *Attention is all you need. Advances in neural information processing systems*, 30.

- Arjovsky, M., Chintala, S. & Bottou, L. (2017). *Wasserstein GAN. In Proceedings of the 31st International Conference on Neural Information Processing System (NIPS).*

- Brock, A., Donahue, J., & Simonyan, K. (2018). *BigGANs: Large Scale GAN Training for High Fidelity Natural Image Synthesis.* https://arxiv.org/abs/1809.11096.

- Karras, T., Aila, T., Laine, S., & Lehtinen, J. (2017). *Progressive Growing of GANs for Improved Quality, Stability, and Variation.* https://arxiv.org/abs/1710.10196.

- Mirza, M., & Osindero, S. (2014). *Conditional Generative Adversarial Nets.* https://arxiv.org/abs/1411.1784.

- Radford, A., Metz, L., & Chintala, S. (2015). *Unsupervised representation learning with deep convolutional generative adversarial networks. 3rd International Conference for Learning Representations.*

- Zhu, J.-Y., Park, T., Isola, P., & Efros, A. A. (2017). *Unpaired Image-to-Image Translation Using Cycle-Consistent Adversarial Networks. In Proceedings of the IEEE International Conference on Computer Vision (ICCV).*

- Ho, J., & Salimans, T. (2022). *Classifier-Free Diffusion Guidance. Advances in Neural Information Processing Systems*, 34.

- Ho, J., Salimans, T., Gritsenko, A. A., Chan, W., Norouzi, M., & Fleet, D. J. (2022). *Video diffusion models.* arXiv preprint arXiv:2205.10477.

- Ho, J., Jain, A., & Abbeel, P. (2020). *Denoising diffusion probabilistic models. Advances in Neural Information Processing Systems*, 33, 6840–6851.

- Luo, S., & Hu, W. (2021). *Diffusion probabilistic models for 3d point cloud generation. Proceedings of the IEEE/CVF Conference on Computer Vision and Pattern Recognition*, 2837–2845.

- Rombach, R., Blattmann, A., Lorenz, D., Esser, P., & Ommer, B. (2021). *High-resolution image synthesis with latent diffusion models. Proceedings of the IEEE/CVF Conference on Computer Vision and Pattern Recognition*, 10684–10695.

- Salimans, T., Karpathy, A., Chen, X., & Kingma, D. P. (2017). *PixelCNN++: Improving the pixelcnn with discretized logistic mixture likelihood and other modifications.* arXiv preprint arXiv:1701.05517.

- Song, Y., Meng, C., & Ermon, S. (2021). *Denoising diffusion implicit models.* arXiv preprint arXiv:2010.02502.

- Song, Y., & Ermon, S. (2021). *Improved techniques for training score-based generative models.* Advances in Neural Information Processing Systems, 33, 12438–12448.

- Sohl-Dickstein, J., Weiss, E. A., Maheswaranathan, N., & Ganguli, S. (2015). *Deep unsupervised learning using nonequilibrium thermodynamics.* arXiv preprint arXiv:1503.03585.

- Ho, J., Jain, A., & Abbeel, P. (2020). *Denoising diffusion probabilistic models. Advances in Neural Information Processing Systems*, 33, 6840–6851.

- Ramesh, A., Pavlov, M., Goh, G., Gray, S., Voss, C., Radford, A., ... & Sutskever, I. (2022). *Zero-shot text-to-image generation. International Conference on Machine Learning*, 8821-8831.

- Brown, T., Mann, B., Ryder, N., Subbiah, M., Kaplan, J. D., Dhariwal, P., ... & Amodei, D. (2020). *Language models are few-shot learners. Advances in neural information processing systems*, 33, 1877–1901.

- Patil, S., Cuenca, P., Lambert, N., & von Platen, P. (2022). *Stable diffusion with diffusers.* Hugging Face Blog. `https://huggingface.co/blog/stable_diffusion`.

- Boris Dayma, Suraj Patil, Pedro Cuenca, Khalid Saifullah, Tanishq Abraham, Phúc Lê, Luke, Ritobrata Ghosh. (2022, June 4). *DALL-E Mini Explained.* W&B; Weights & Biases, Inc. `https://wandb.ai/dalle-mini/dalle-mini/reports/DALL-E-Mini-Explained-with-Demo--Vmlldzo4NjIxODA`.

Tracing the Foundations of Natural Language Processing and the Impact of the Transformer

The transformer architecture is a key advancement that underpins most modern generative language models. Since its introduction in 2017, it has become a fundamental part of **natural language processing** (**NLP**), enabling models such as **Generative Pre-trained Transformer 4** (**GPT-4**) and Claude to advance text generation capabilities significantly. A deep understanding of the transformer architecture is crucial for grasping the mechanics of modern **large language models** (**LLMs**).

In the previous chapter, we explored generative modeling techniques, including **generative adversarial networks** (**GANs**), diffusion models, and **autoregressive** (**AR**) transformers. We discussed how Transformers can be leveraged to generate images from text. However, transformers are more than just one generative approach among many; they form the basis for nearly all state-of-the-art generative language models.

In this chapter, we'll cover the evolution of NLP that ultimately led to the advent of the transformer architecture. We cannot cover all the critical steps forward, but we will attempt to cover major milestones, starting with early linguistic analysis techniques and statistical language modeling, followed by advancements in **recurrent neural networks** (**RNNs**) and **convolutional neural networks** (**CNNs**) that highlight the potential of **deep learning** (**DL**) for NLP. Our main objective will be to introduce the transformer—its basis in DL, its self-attention architecture, and its rapid evolution, which has led to LLMs and this phenomenon we call **generative AI** (**GenAI**).

Understanding the origins and mechanics of the transformer architecture is important for recognizing its groundbreaking impact. The principles and modeling capabilities introduced by transformers are carried forward by all modern language models built upon this framework. We will build our intuition for Transformers through historical context and hands-on implementation, as this foundational understanding is key to understanding the future of GenAI.

Early approaches in NLP

Before the widespread use of **neural networks** (**NNs**) in language processing, NLP was largely grounded in methods that counted words. Two particularly notable techniques were **count vectors** and **Term Frequency-Inverse Document Frequency** (**TF-IDF**). In essence, count vectors tallied up how often each word appeared in a document. Building on this, Dadgar et al. applied the TF-IDF algorithm (historically used for information retrieval) to text classification in 2016. This method assigned weights to words based on their significance in one document relative to their occurrence across a collection of documents. These count-based methods were successful for tasks such as searching and categorizing. However, they presented a key limitation in that they could not capture the semantic relationships between words, meaning they could not interpret the nuanced meanings of words in context. This challenge paved the way for exploring NNs, offering a deeper and more nuanced way to understand and represent text.

Advent of neural language models

In 2003, Yoshua Bengio's team at the University of Montreal introduced the **Neural Network Language Model** (**NNLM**), a novel approach to language technology. The NNLM was designed to predict the next word in a sequence based on prior words using a particular type of **neural network** (**NN**). The design prominently featured hidden layers that learned word embeddings, which are compact vector representations capturing the core semantic meanings of words. This aspect was absent in count-based approaches. However, the NNLM was still limited in its ability to interpret longer sequences and handle large vocabularies. Despite these limitations, the NNLM sparked widespread exploration of NNs in language modeling.

The introduction of the NNLM highlighted the potential of NNs in language processing, particularly using word embeddings. Yet, its limitations with long sequences and large vocabulary signaled the need for further research.

Distributed representations

Following the inception of the NNLM, NLP research was propelled toward crafting high-quality word vector representations. These representations could be initially learned from extensive sets of unlabeled text data and later applied to downstream models for various tasks. The period saw the emergence of two prominent methods: **Word2Vec** (introduced by Mikolov et al., 2013) and **Global Vectors** (**GloVe**, introduced by Pennington et al., 2014). These methods applied **distributed representation** to craft high-quality word vector representations. Distributed representation portrays items such as words not as unique identifiers but as sets of continuous values or vectors. In these vectors, each value corresponds to a specific feature or characteristic of the item. Unlike traditional representations, where each item has a unique symbol, distributed representations allow these items to share features with others, enabling a more intelligent capture of underlying patterns in the data.

Let us elucidate this concept a bit further. Suppose we represent words based on two features: **Formality** and **Positivity**. We might have vectors such as the following:

```
Formal: [1, 0]

Happy: [0, 1]

Cheerful: [0, 1]
```

In this example, each element in the vector corresponds to one of these features. In the vector for `Formal`, the 1 element under *Formality* indicates that the word is formal, while the 0 element under *Positivity* indicates neutrality in terms of positivity. Similarly, for `Happy` and `Cheerful`, the 1 element under *Positivity* indicates that these words have a positive connotation. This way, distributed representation captures the essence of words through vectors, allowing for shared features among different words to understand underlying patterns in data.

Word2Vec employs a relatively straightforward approach where NNs are used to predict the surrounding words for each target word in a dataset. Through this process, the NN ascertains values or "weights" for each target word. These weights form a vector for each word in a **continuous vector space**—a mathematical space wherein each point represents a possible value a vector can take. In the context of NLP, each dimension of this space corresponds to a feature, and the position of a word in this space captures its semantic or linguistic relationships to other words.

These vectors form a **feature-based representation**—a type of representation where each dimension represents a different feature that contributes to the word's meaning. Unlike a symbolic representation, where each word is represented as a unique symbol, a feature-based representation captures the semantic essence of words in terms of shared features.

On the other hand, GloVe adopts a different approach. It analyzes the **global co-occurrence statistics**—a count of how often words appear together in a large text corpus. GloVe learns vector representations that capture the relationships between words by analyzing these counts across the entire corpus. This method also results in a distributed representation of words in a continuous vector space, capturing **semantic similarity**—a measure of the degree to which two words are similar in meaning. In a continuous vector space, we can think about semantic similarity as the simple geometric proximity of vectors representing words.

To further illustrate, suppose we have a tiny corpus of text containing the following sentences:

"Coffee is hot."

"Ice cream is cold."

From this corpus, GloVe would notice that "*coffee*" co-occurs with "*hot*" and "*ice cream*" co-occurs with "*cold*." Through its optimization process, it would aim to create vectors for these words in a way that reflects these relationships. In this oversimplified example, GloVe might produce a vector such as this:

```
Coffee: [1, 0]

Hot: [0.9, 0]

Ice Cream: [0, 1]

Cold: [0, 0.9]
```

Here, the closeness of the vectors for "*coffee*" and "*hot*" (and, similarly, "*ice cream*" and "*cold*") in this space reflects the co-occurrence relationships observed in the corpus. The vector difference between "*coffee*" and "*hot*" might resemble the vector difference between "*ice cream*" and "*cold*," capturing the contrasting temperature relationships in a geometric way within the vector space.

Both Word2Vec and GloVe excel at encapsulating relevant semantic information about words to represent an efficient **encoding**—a compact way of representing information that captures the essential features necessary for a task while reducing the dimensionality and complexity of the data.

These methodologies in creating meaningful vector representations served as a step toward the adoption of **transfer learning** in NLP. The vectors provide a shared semantic foundation that facilitates the transfer of learned relationships across varying tasks.

Transfer Learning

GloVe and other methods of deriving distributed representations paved the way for transfer learning in NLP. By creating rich vector representations of words that encapsulate semantic relationships, these methods provided a foundational understanding of text. The vectors serve as a shared base of knowledge that can be applied to different tasks. When a model, initially trained on one task, is utilized for another, the pre-learned vector representations aid in preserving the semantic understanding, thereby reducing the data or training needed for the new task. This practice of transferring acquired knowledge has become fundamental for efficiently addressing a range of NLP tasks.

Consider a model trained to understand sentiments (positive or negative) in movie reviews. Through training, this model has learned distributed representations of words, capturing sentiment-related nuances. Now, suppose there is a new task: understanding sentiments in product reviews. Instead of training a new model from the beginning, transfer learning allows us to use the distributed representations from the movie review task to initiate training for the product review task. This could lead to quicker training and better performance, especially with limited data for the product review task.

The effectiveness of transfer learning, bolstered by distributed representations from methods such as GloVe, highlighted the potential of leveraging pre-existing knowledge for new tasks. It was a precursor to the integration of NNs in NLP, highlighting the benefits of utilizing learned representations across tasks. The advent of NNs in NLP brought about models capable of learning even richer representations, further amplifying the impact and scope of transfer learning.

Advent of NNs in NLP

The advent of NNs in NLP marked a monumental shift in the field's capability to understand and process language. Building upon the groundwork laid by methodologies such as Word2Vec, GloVe, and the practice of transfer learning, NNs introduced a higher level of abstraction and learning capacity. Unlike previous methods that often relied on hand-crafted features, NNs could automatically learn intricate patterns and relationships from data. This ability to learn from data propelled NLP into a new era where models could achieve unprecedented levels of performance across a myriad of language-related tasks. The emergence of architectures such as CNNs and RNNs, followed by the revolutionary transformer architecture, showcased the remarkable versatility and efficacy of NNs in tackling complex NLP challenges. This transition not only accelerated the pace of innovation but also expanded the horizon of what could be achieved in understanding human language computationally.

Language modeling with RNNs

Despite how well these distributed word vectors excelled at encoding local semantic relationships, modeling long-range dependencies would require a more sophisticated network architecture. This led to the use of RNNs. RNNs (originally introduced by Elman in 1990) are a type of NN architecture that processes data sequences by iterating through each element of the sequence while maintaining a dynamic internal state that captures information about the previous elements. Unlike traditional **feedforward networks** (**FNNs**) that processed each input independently, RNNs introduced iterations that allowed information to be passed from one step in the sequence to the next, enabling them to capture temporal dependencies in data. The iterative processing and dynamic updating in NNs enable them to learn and represent relationships within the text. These networks can capture contextual connections and interdependencies across sentences or even entire documents.

However, standard RNNs had technical limitations when dealing with long sequences. This led to the development of **long short-term memory** (**LSTM**) networks. LSTMs were first introduced by Hochreiter and Schmidhuber in 1997. They were a special class of RNNs designed to address the **vanishing gradient** problem, which is the challenge where the network cannot learn from earlier parts of a sequence as the sequence gets longer. LSTMs applied a unique gating architecture to control the flow of information within the network, enabling them to maintain and access information over long sequences without suffering from the vanishing gradient problem.

The name "**long short-term memory**" refers to the network's ability to keep track of information over both short and long sequences of data:

- **Short-term**: LSTMs can remember recent information, which is useful for understanding the current context. For example, in language modeling, knowing the last few words can be crucial for predicting the next word. Consider a phrase such as, "The cat, which already ate a lot, was not hungry." As the LSTM processes the text, when it reaches the word "not," the recent information that the cat "ate a lot" is crucial to predict the next word, "hungry," accurately.

- **Long-term**: Unlike standard RNNs, LSTMs are also capable of retaining information from many steps back in the sequence, which is particularly useful for long-range dependencies, where a piece of information early in a sentence could be important for understanding a word much later in the sequence. In the same phrase, the information that "The cat" is the subject of the sentence is introduced early on. This information is crucial later to understand who "was not hungry" as it processes the later part of the sentence.

The **M** or **memory** in LSTMs is maintained through a unique architecture that employs three gating mechanisms—input, output, and forget gates. These gates control the flow of information within the network, deciding what information should be kept, discarded, or used at each step in the sequence, enabling LSTMs to maintain and access information over long sequences. Effectively, these gates and the network state allowed LTSMs to carry the "memory" across time steps, ensuring that valuable information was retained throughout the processing of the sequence.

Ultimately, LSTMs obtained state-of-the-art results on many language modeling and text classification benchmarks. They became the dominant NN architecture for NLP tasks due to their ability to capture short- and long-range contextual relationships.

The success of LSTMs demonstrated the potential of neural architectures in capturing the complex relationships inherent in language, significantly advancing the field of NLP. However, the continuous pursuit of more efficient and effective models led the community toward exploring other NN architectures.

Rise of CNNs

Around 2014, the NLP domain witnessed a rise in the popularity of CNNs for tackling NLP tasks, a notable shift led by Yoon Kim. CNNs (originally brought forward by LeCun et al. for image recognition) operate based on convolutional layers that scan the input by moving a filter (or kernel) across the input data, at each position calculating the dot product of the filter's weights and the input data. In NLP, these layers work over local n-gram windows (consecutive sequences of *n* words) to identify patterns or features, such as specific sequences of words or characters in the text. Employing convolutional layers over local n-gram windows, CNNs scan and analyze the data to detect initial patterns or features. Following this, pooling layers are used to reduce the dimensionality of the data, which helps in both reducing computational complexity and focusing on the most salient features identified by the convolutional layers.

Combining convolutional and pooling layers, CNNs can extract hierarchical features. These features represent information at different levels of abstraction by combining simpler, lower-level features to form more complex, higher-level features. In NLP, this process might start with detecting basic patterns such as common word pairs or phrases in the initial layers, progressing to recognizing more abstract concepts such as semantic relationships in the higher layers.

For comparison, we again consider a scenario where a CNN is employed to analyze and categorize customer reviews into positive, negative, or neutral sentiments:

- **Lower-level features (initial layers)**: The CNN might identify basic patterns such as common word pairs or phrases in the initial layers. For instance, it might recognize phrases such as "great service," "terrible experience," or "not happy."

- **Intermediate-level features (middle layers)**: As data progresses through the network, middle layers might start recognizing more complex patterns, such as negations ("not good") or contrasts ("good but expensive").

- **Higher-level features (later layers)**: The CNN could identify abstract concepts such as overall sentiment in the later layers. For instance, it might deduce a positive sentiment from phrases such as "excellent service" or "loved the ambiance" and a negative sentiment from phrases such as "worst experience" or "terrible food."

In this way, CNNs inherently learn higher-level abstract representations of text. Although they lack the sequential processing characteristic of RNNs, they offer a computational advantage due to their inherent **parallelism** or ability to process multiple parts of the data simultaneously. Unlike RNNs, which process sequences iteratively and require the previous step to be completed before proceeding to the next, CNNs can process various parts of the input data in parallel, significantly speeding up training times.

CNNs, while efficient, have a limitation in their convolution operation, which only processes local data from smaller or nearby regions, thereby missing relationships across more significant portions of the entire input data, referred to as global information. This gave rise to attention-augmented convolutional networks that integrate self-attention with convolutions to address this limitation. Self-attention, initially used in sequence and generative modeling, was adapted for visual tasks such as image classification, enabling the network to process and capture relationships across the entire input data. However, attention augmentation, which combines convolutions and self-attention, yielded the best results. This method retained the computational efficiency of CNNs and captured global information, marking an advancement in image classification and object detection tasks. We will discuss self-attention in detail later as it became a critical component of the transformer.

The ability of CNNs to process multiple parts of data simultaneously marked a significant advancement in computational efficiency, paving the way for further innovations in NN architectures for NLP. As the field progressed, a pivotal shift occurred with the advent of attention-augmented NNs, introducing a new paradigm in how models handle sequential data.

The emergence of the Transformer in advanced language models

In 2017, inspired by the capabilities of CNNs and the innovative application of attention mechanisms, Vaswani et al. introduced the transformer architecture in the seminal paper *Attention is All You Need*. The original transformer applied several novel methods, particularly emphasizing the instrumental impact of attention. It employed a **self-attention mechanism**, allowing each element in the input sequence to focus on distinct parts of the sequence, capturing dependencies regardless of their positions in a structured manner. The term "self" in "self-attention" refers to how the attention mechanism is applied to the input sequence itself, meaning each element in the sequence is compared to every other element to determine its attention scores.

To truly appreciate how the transformer architecture works, we can describe how the components in its architecture play a role in handling a particular task. Suppose we need our transformer to translate the English sentence "*Hello, how are you?*" into French: "*Bonjour, comment ça va?*" Let us walk through this step by step to examine and elucidate how the transformer might accomplish this task. For now, we will describe each step in detail and later implement the full architecture using Python.

Components of the transformer architecture

Before diving into how the transformer model fulfills our translation task, we need to understand the steps involved. The complete architecture is quite dense, so we will break it down into small, logical, and digestible components.

First, we discuss the two components central to the architectural design of the transformer model: the encoder and decoder stacks. We will also explain how data flows within these layer stacks, including the concept of tokens, and how relationships between tokens are captured and refined using critical techniques such as self-attention and FFNs.

Then, we transition into the training process of the transformer model. Here, we review fundamental concepts such as batches, masking, the training loop, data preparation, optimizer selection, and strategies to improve performance. We will explain how the transformer optimizes performance using a loss function, which is crucial in shaping how the model learns to translate.

Following the training process, we discuss model inference, which is how our trained model generates translations. This section points out the order in which individual model components operate during translation and emphasizes the importance of each step.

As discussed, central to the transformer are two vital components, often called the encoder stack and the decoder stack.

Encoder and decoder stacks

In the context of the transformer model, **stacks** reference a hierarchical arrangement of **layers**. Each layer in this context is, in fact, an NN layer like the layers we come across in classical DL models. While a layer is a level in the model where specific computational operations occur, a stack refers to multiple such layers arranged consecutively.

Encoder stack

Consider our example sentence "*Hello, how are you?*". We first convert it into tokens. Each token typically represents a word. In the case of our example sentence, tokenization would break it down into separate tokens, resulting in the following:

```
["Hello", ",", "how", "are", "you", "?"]
```

Here, each word or punctuation represents a distinct token. These tokens are then transformed into numerical representations, also known as **embeddings**. These embedding vectors capture the semantic meaning and context of the words, enabling the model to understand and process the input data effectively. The embeddings aid in capturing complex relationships and contexts from the original English input sentence through this series of transformations across layers.

This stack comprises multiple layers, where each layer applies self-attention and FFN computations on its input data (which we will describe in detail shortly). The embeddings iteratively capture complex relationships and context from the original English input sentence through this series of transformations across layers.

Decoder stack

Once the encoder completes its task, the output vectors—or the embeddings of the input sentence that hold its contextual information—are passed on to the decoder. Within the decoder stack, multiple layers work sequentially to generate a French translation from the embeddings.

The process begins by converting the first embedding into the French phrase "*Bonjour.*" The subsequent layer uses the following embedding and context from the previously generated words to predict the next word in the French sentence. This process is repeated through all the layers in the stack, each using input embeddings and generated words to define and refine the translation.

The decoder stack progressively builds (or decodes) the translated sentence through this iterative process, arriving at "*Bonjour, comment ça va?*".

With an overall understanding of the encoder-decoder structure, our next step is unraveling the intricate operations within each stack. However, before delving into the self-attention mechanism and FFNs, there is one vital component we need to understand — positional encoding. Positional encoding is paramount to the transformer's performance because it gives the transformer model a sense of the order of words, something subsequent operations in the stack lack.

Positional encoding

Every word in a sentence holds two types of information — its meaning and its role in the larger context of the sentence. The contextual role often stems from a word's position in the arrangement of words. A sentence such as "*Hello, how are you?*" makes sense because the words are in a specific order. Change that to "*Are you, how hello?*" and the meaning becomes unclear.

Consequently, Vaswani et al. introduced **positional encoding** to ensure that the transformer encodes each word with additional data about its position in the sentence. Positional encodings are computed using a blend of sine and cosine functions across different frequencies, which generate a unique set of values for each position in a sequence. These values are then added to the original embeddings of the tokens, providing a way for the model to capture the order of words. These enriched embeddings are then ready to be processed by the self-attention mechanism in the subsequent layers of the transformer model.

Self-attention mechanism

As each token of our input sentence *"Hello, how are you?"* passes through each layer of the encoder stack, it undergoes a transformation via the self-attention mechanism.

As the name suggests, the self-attention mechanism allows each token (word) to attend to (or focus on) other vital tokens to understand the full context within the sentence. Before encoding a particular word, this attention mechanism interprets the relationship between each word and the others in the sequence. It then assigns distinct attention scores to different words based on their relevance to the current word being processed.

Consider again our input sentence *"Hello, how are you?"*. When the self-attention mechanism is processing the last word, *"you,"* it does not just focus on *"you."* Instead, it takes into consideration the entire sentence: it looks at *"Hello,"* glances over *"how,"* reflects on *"are,"* and, of course, focuses on *"you."*

In doing so, it assigns various levels of attention to each word. You can visualize attention (*Figure 3.1*) as lines connecting *"you"* to every other word. The line to *"Hello"* might be thick, indicating a lot of attention, representing the influence of *"Hello"* on the encoding of *"you."* The line connecting *"you"* and *"how"* might be thinner, suggesting less attention given to *"how."* The lines to *"are"* and *"you"* would have other thicknesses based on how they help in providing context to *"you"*:

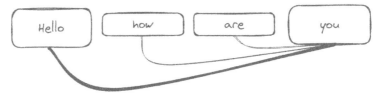

Figure 3.1: Self-attention mechanism

This way, when encoding *"you,"* a weighted mix of the entire sentence is considered, not just the single word. And these weights defining the mix are what we refer to as attention scores.

The self-attention mechanism is implemented through a few steps:

1. Initially, each input word is represented as a vector, which we obtain from the word embedding.
2. These vectors are then mapped to new vectors called query, key, and value vectors through learned transformations.
3. An attention score for each word is then computed by taking the dot product of the query vector of the word with the key vector of every other word, followed by a SoftMax operation (which we will describe later).
4. These scores indicate how much focus to place on other parts of the input sentence for each word as it is encoded.
5. Finally, a weighted sum of the value vectors is computed based on these scores to give us our final output vectors, or the self-attention outputs.

It is important to note that this computation is done for each word in the sentence. This ensures a comprehensive understanding of the context in the sentence, considering multiple parts of the sentence at once. This concept set the transformer apart from nearly every model that came before it.

Instead of running the self-attention mechanism once (or "single-head" attention), the transformer replicates the self-attention mechanism multiple times in parallel. Each replica or head operates on the same input but has its own independent set of learned parameters to compute the attention scores. This allows each head to learn different contextual relationships between words. This parallel process is known as **multi-head attention** (**MHA**).

Imagine our sentence "*Hello, how are you?*" again. One head might concentrate on how "*Hello*" relates to "*you*," whereas another head might focus more on how "*how*" relates to "*you*." Each head has its own set of query, key, and value weights, further enabling them to specialize and learn different things. The outputs of these multiple heads are then concatenated and transformed to produce final values passed onto the next layer in the stack.

This multi-head approach allows the model to capture a wider range of information from the same input words. It is like having several perspectives on the same sentence, each providing unique insights.

So far, for our input sentence "*Hello, how are you?*", we have converted each word into token representations, which are then contextualized using the MHA mechanism. Through parallel self-attention, our transformer can consider the full range of interactions between each word and every other word in the sentence. We now have a set of diverse and context-enriched word representations, each containing a textured understanding of a word's role in the sentence. However, this contextual understanding contained within the attention mechanism is just one component of the information processing in our transformer model. Next comes another layer of interpretation through position-wise FFNs. The FFN will add further nuances to these representations, making them more informative and valuable for our translation task.

In the next section, we discuss a vital aspect of the transformer's training sequence: masking. Specifically, the transformer applies causal (or look-ahead) masking during the decoder self-attention to ensure that each output token prediction depends only on previously generated tokens, not future unknown tokens.

Masking

The transformer applies two types of masking during training. The first is a preprocessing step to ensure input sentences are of the same length, which enables efficient batch computation. The second is look-ahead (or causal) masking, which allows the model to selectively ignore future tokens in a sequence. This type of masking occurs in the self-attention mechanism in the decoder and prevents the model from peeking ahead at future tokens in the sequence. For example, when translating the word "*Hello*" to French, look-ahead masking ensures that the model does not have access to the subsequent words "*how*," "*are*," or "*you*." This way, the model learns to generate translations based on the current and preceding words, adhering to a natural progression in translation tasks, mimicking that of human translation.

With a clearer understanding of how data is prepared and masked for training, we now transition to another significant aspect of the training process: hyperparameters. Unlike parameters learned from the data, hyperparameters are configurations set before training to control the model optimization process and guide the learning journey. The following section will explore various hyperparameters and their roles during training.

SoftMax

To understand the role of the FFN, we can describe its two primary components—linear transformations and an activation function:

- **Linear transformations** are essentially matrix multiplications. Think of them as tools that reshape or tweak the input data. In the FFN, these transformations occur twice, where two different weights (or matrices) are used.

- A **rectified linear unit** (**ReLU**) function is applied between these two transformations. The role of the ReLU function is to introduce non-linearity in the model. Simply put, the ReLU function allows the model to capture patterns within the input data that are not strictly proportional, i.e., non-linear, which is typical of **natural language** (**NL**) data.

The FFN is called **position-wise** because it treats each word in the sentence separately (position by position), regardless of the sequence. This contrasts with the self-attention mechanism, which considers the entire sequence at once.

So, let us attempt to visualize the process: Imagine our word "*Hello*" arriving here after going through the self-attention mechanism. It carries with it information about its own identity mixed with contextual references to "*how,*" "*are,*" and "*you.*" This integrated information resides within a vector that characterizes "*Hello.*"

When "*Hello*" enters the FFN, picture it as a tunnel with two gates. At the first gate (or linear layer), "*Hello*" is transformed by a matrix multiplication operation, changing its representation. Afterward, it encounters the ReLU function—which makes the representation non-linear.

After this, "*Hello*" passes through a second gate (another linear layer), emerging on the other side transformed yet again. The core identity of "*Hello*" remains but is now imbued with even more context, carefully calibrated and adjusted by the FFN.

Once the input passes through the gates, there is one additional step. The transformed vector still must be converted into a form that can be interpreted as a prediction for our final translation task.

This brings us to using the SoftMax function, the final transformation within the transformer's decoder. After the vectors pass through the FFN, they are further processed through a final linear layer. The result is then fed into a SoftMax function.

SoftMax serves as a mechanism for converting the output of our model into a form that can be interpreted as probabilities. In essence, the SoftMax function will take the output from our final linear layer (which could be any set of real numbers) and transform it into a distribution of probabilities, representing the likelihood of each word being the next word in our output sequence. For example, if our target vocabulary includes "*Bonjour*," "*Hola*," "*Hello*," and "*Hallo*," the SoftMax function will assign each of these words a probability, and the word with the highest probability will be chosen as the output translation for the word "*Hello*." We can illustrate with this oversimplified representation of the output probabilities:

```
[ Bonjour: 0.4, Hola: 0.3, Hello: 0.2, Hallo: 0.1 ]
```

Figure 3.2 shows a more complete (albeit oversimplified) view of the flow of information through the architecture.

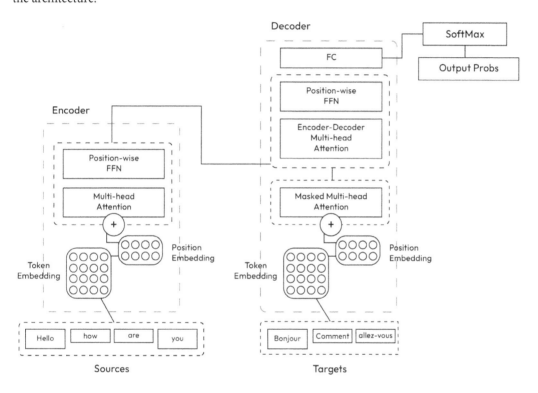

Figure 3.2: A simplified illustration of the transformer

Now that we've introduced the architectural components of the transformer, we are poised to understand how its components work together.

Sequence-to-sequence learning

The components of a transformer come together to learn from data using a mechanism known as **sequence-to-sequence (Seq2Seq)** learning, a subset of **supervised learning (SL)**. Recall that SL is a technique that uses labeled data to train models to predict outcomes accurately. In Seq2Seq learning, we provide the transformer with training data that comprises examples of input and corresponding correct output, which, in this case, are correct translations. Seq2Seq learning is particularly well suited for tasks such as machine translation where both the input and output are sequences of words.

The very first step in the learning process is to convert each word in the phrase into tokens, which are then transformed into numerical embeddings. These embeddings carry the semantic essence of each word. Positional encodings are computed and added to these embeddings to imbue them with positional awareness.

As these enriched embeddings traverse through the encoder stack, within each layer, the self-attention mechanism refines the embeddings by aggregating contextual information from the entire phrase. Following self-attention, each word's embedding undergoes further transformation in the position-wise FFNs, adjusting the embeddings to capture even more complex relationships.

Upon exiting the encoder, the embeddings now hold a rich mixture of semantic and contextual information. They are passed onto the decoder stack, which aims to translate the phrase into another language (that is, the target sequence). As with the encoder, each layer in the decoder also employs self-attention and position-wise FFNs, but with an additional layer of cross-attention that interacts with the encoder's outputs. This interaction helps align the input and output phrases, a crucial aspect of translation.

As the embeddings move through the decoder layers, they are progressively refined to represent the translated phrase that the model will predict. The final layer of the decoder processes the embeddings through a linear transformation and SoftMax function to produce a probability distribution over the target vocabulary. This distribution defines the model's predicted likelihood for each potential next token at each step. The decoder then samples from this distribution to select the token with the highest predicted probability as its next output. By iteratively sampling the most likely next tokens according to the predicted distributions, the decoder can autoregressively generate the full translated output sequence token by token.

However, for the transformer to reliably sample from the predicted next-token distributions to generate high-quality translations, it must progressively learn by iterating over thousands of examples of input-output pairs. In the next section, we explore model training in further detail.

Model training

As discussed, the primary goal of the training phase is to refine the model's parameters to facilitate accurate translation from one language to another. But what does the refinement of parameters entail, and why is it pivotal?

Parameters are internal variables that the model utilizes to generate translations. Initially, these parameters are assigned random values, which are adjusted with each training iteration. Again, the model is provided with training data that comprises thousands of examples of input data and corresponding correct output, which, in this case, is the correct translation. It then compares its predicted output tokens to the correct (or actual) target sequences using an error (or loss) function.

Based on the loss, the model updates its parameters, gradually improving its ability to choose the correct item in the sequence at each step of decoding. This slowly refines the probability distributions.

Over thousands of training iterations, the model learns associations between source and target languages. Eventually, it acquires enough knowledge to decode coherent, human-like translations from unseen inputs by relying on patterns discovered during training. Therefore, training drives the model's ability to produce accurate target sequences from the predicted vocabulary distributions.

After training on sufficient translation pairs, the transformer reaches reliable translation performance. The trained model can then take in new input sequences and output translated sequences by generalizing to that new data.

For instance, with our example sentence "*Hello, how are you?*" and its French translation "*Bonjour, comment ça va?*", the English sentence serves as the input, and the French sentence serves as the target output. The training data comprises many translated pairs. Each time the model processes a batch of data, it generates predictions for the translation, compares them to the actual target translations, and then adjusts its parameters to reduce the discrepancy (or minimize the loss) between the predicted and actual translations. This is repeated with numerous batches of data until the model's translations are sufficiently accurate.

Hyperparameters

Again, unlike parameters, which the model learns from the training data, hyperparameters are preset configurations that govern the training process and the structure of the model. They are a crucial part of setting up a successful training run.

Some key hyperparameters in the context of transformer models include the following:

- **Learning rate**: This value determines the step size at which the optimizer updates the model parameters. A higher learning rate could speed up the training but may overshoot the optimal solution. A lower learning rate may result in a more precise convergence to the optimal solution, albeit at the cost of longer training time. We will discuss optimizers in detail in the next section.

- **Batch size**: The number of data examples processed in a single batch affects the computational accuracy and the memory requirements during training.

- **Model dimensions**: The model's size (for example, the number of layers in the encoder and decoder, the dimensionality of the embeddings, and so on) is a crucial hyperparameter that affects the model's capacity to learn and generalize.

- **Optimizer settings**: Choosing an optimizer and its settings, such as the initial learning rate, beta values in the Adam optimizer, and so on, are also considered hyperparameters. Again, we will explore optimizers further in the next section.

- **Regularization terms**: Regularization terms such as dropout rate are hyperparameters that help prevent overfitting by adding some form of randomness or constraint to the training process.

Selecting the proper values for hyperparameters is crucial for the training process as it significantly impacts the model's performance and efficiency. It often involves hyperparameter tuning, which involves experimentation and refining to find values for hyperparameters that yield reliable performance for a given task. Hyperparameter tuning can be somewhat of an art and a science. We will touch on this more in later chapters.

With a high-level grasp of hyperparameters, we will move on to the choice of optimizer, which is pivotal in controlling how efficiently the model learns from the training data.

Choice of optimizer

The optimizer is a fundamental component of the training process and is responsible for updating the model's parameters to minimize error. Different optimizers have different strategies for navigating the parameter space to find a set of parameter values that yield low loss (or less error). The choice of optimizer can significantly impact the speed and quality of the training process.

In the context of transformer models, the Adam optimizer is often the optimizer of choice due to its efficiency and empirical success in training deep networks. Adam adapts learning rates during training. For simplicity, we will not explore all the possible optimizers but instead describe their purpose.

The optimizer's primary task is to fine-tune the model's parameters to reduce translation errors, progressively guiding the model toward the desired level of performance. However, an over-zealous optimization could lead the model to memorize the training data, failing to generalize well to unseen data. To mitigate this, we employ regularization techniques.

In the next section, we will explore regularization—a technique that works with optimization to ensure that while the model learns to minimize translation errors, it also remains adaptable to new, unseen data.

Regularization

Regularization techniques are employed to deter the model from memorizing the training data (a phenomenon known as overfitting) and to promote better performance on new, unseen data. Overfitting arises when the model, to minimize the error, learns the training data to such an extent that it captures useless patterns (or noise) along with the actual patterns. This over-precision in learning the training data leads to a decline in performance when the model is exposed to new data.

Let us revisit our simple scenario where we train a model to translate English greetings to French greetings using a dataset that includes the word "*Hello*" and its translation "*Bonjour*." If the model is overfitting, it may memorize the exact phrases from the training data without understanding the broader translation pattern.

In an overfit scenario, suppose the model learns to translate "*Hello*" to "*Bonjour*" with a probability of 1.0 because that is what it encountered most often in the training data. When presented with new, unseen data, it may encounter variations it has not seen before, such as "*Hi*," which should also translate to "*Bonjour*." However, due to overfitting, the model might fail to generalize from "*Hello*" to "*Hi*" as it is overly focused on the exact mappings it saw during training.

Several regularization techniques can mitigate the overfitting problem. These techniques apply certain constraints on the model's parameters during training, encouraging the model to learn a more generalized representation of the data rather than memorizing the training dataset.

Here are some standard regularization techniques used in the context of transformer models:

- **Dropout**: In the context of NN-based models such as the transformer, the term "neurons" refers to individual elements within the model that work together to learn from the data and make predictions. Each neuron learns specific aspects or features from the data, enabling the model to understand and translate text. During training, dropout randomly deactivates or "drops out" a fraction of these neurons, temporarily removing them from the network. This random deactivation encourages the model to spread its learning across many neurons rather than relying too heavily on a few. By doing so, dropout helps the model to better generalize its learning to unseen data rather than merely memorizing the training data (that is, overfitting).

- **Layer normalization**: Layer normalization is a technique that normalizes the activations of neurons in a layer for each training example rather than across a batch of examples. This normalization helps stabilize the training process and acts as a form of regularization, preventing overfitting.

- **L1 or L2 regularization**: L1 regularization, also known as Lasso, adds a penalty equal to the absolute magnitude of coefficients, promoting parameter sparsity. L2 regularization, or Ridge, adds a penalty based on the square of the coefficients, discouraging large values to prevent overfitting. Although these techniques help in controlling model complexity and enhancing generalization, they were not part of the transformer's initial design.

By employing these regularization techniques, the model is guided toward learning more generalized patterns in the data, which improves its ability to perform well on unseen data, thus making the model more reliable and robust in translating new text inputs.

Throughout the training process, we have mentioned the loss function and discussed how the optimizer leverages it to adjust the model's parameters, aiming to minimize prediction error. The loss function quantifies the model's performance. We discussed how regularization penalizes the loss function to prevent overfitting, encouraging the model to learn simpler, more generalizable patterns. In the next section, we look closer at the nuanced role of the loss function itself.

Loss function

The loss function is vital in training the transformer model, quantifying the differences between the model's predictions and the actual data. In language translation, this error is measured between generated and actual translations in the training dataset. A common choice for this task is cross-entropy loss, which measures the difference between the model's predicted probability distribution across the target vocabulary and the actual distribution, where the truth has a probability of 1 for the correct word and 0 for the rest.

The transformer often employs a variant known as label-smoothed cross-entropy loss. Label smoothing adjusts the target probability distribution during training, slightly lowering the probability for the correct class and increasing the probability for all other classes, which helps prevent the model from becoming too confident in its predictions. For instance, with a target vocabulary comprising *"Bonjour," "Hola,"* *"Hello,"* and *"Hallo,"* and assuming *"Bonjour"* is the correct translation, a standard cross-entropy loss would aim for the probability distribution of Bonjour: 1.0, Hola: 0.0, Hello: 0.0, Hallo: 0.0. However, the label-smoothed cross-entropy loss would slightly adjust these probabilities, as follows:

```
[ "Bonjour": 0.925, "Hola": 0.025, "Hello": 0.025, "Hallo": 0.025 ]
```

The smoothing reduces the model's confidence and promotes better generalization to unseen data. With a clearer understanding of the loss function's role, we can move on to the inference phase, where the trained model generates translations for new, unseen data.

Inference

Having traversed the training landscape, our trained model is now adept with optimized parameters to tackle the translation task. In the inference stage, these learned parameters are employed to translate new, unseen text. We will continue with our example phrase *"Hello, how are you?"* to elucidate this process.

The inference stage is the practical application of the trained model on new data. The trained parameters, refined after numerous iterations during training, are now used to translate text from one language to another. The inference steps can be described as follows:

1. **Input preparation**: Initially, our phrase "Hello, how are you?" is tokenized and encoded into a format that the model can process, akin to the preparation steps in the training phase.

2. **Passing through the model**: The encoded input is then propagated through the model. As it navigates through the encoder and decoder stacks, the trained parameters guide the transformation of the input data, inching closer to accurate translations at each step.

3. **Output generation**: At the culmination of the decoder stack, the model generates a probability distribution across the target vocabulary for each word in the input text. For the word "Hello," a probability distribution is formed over the target vocabulary, which, in our case, comprises French words. The word with the highest probability is selected as the translation. This process is replicated for each word in the phrase, rendering the translated output *"Bonjour, comment ça va?"*.

Now that we understand how the model produces the final output, we can implement a transformer model step by step to solidify the concepts we have discussed. However, before we dive into the code, we can briefly give a synopsis of the end-to-end architecture flow:

1. **Input tokenization**: The initial English phrase *"Hello, how are you?"* is tokenized into smaller units such as *"Hello," ",", "how,"* and so on.

2. **Embeddings**: These tokens are then mapped to continuous vector representations through an embedding layer.

3. **Positional encoding**: To preserve the order of the sequence, positional encodings are added to the embeddings.

4. **Encoder self-attention**: The embedded input sequence navigates through the encoder's sequence of self-attention layers. Here, each word gauges the relevance of every other word to comprehend the full context.

5. **FFN**: The representations are subsequently refined by position-wise FFNs within each encoder layer.

6. **Encoder output**: The encoder renders contextual representations capturing the essence of the input sequence.

7. **Decoder attention**: Incrementally, the decoder crafts the output sequence, employing self-attention solely on preceding words to maintain the sequence order.

8. **Encoder-decoder attention**: The decoder evaluates the encoder's output, centering on pertinent input context while generating each word in the output sequence.

9. **Output layers**: The decoder feeds its output to the linear and SoftMax layers to produce "*Bonjour, comment ça va?*

At the end of this chapter, we will adapt a best-in-class implementation of the original transformer (Huang et al., 2022) into a minimal example that could later be trained on various downstream tasks. This will serve as a theoretical exercise to further solidify our understanding. In practice, we would rely on pre-trained or foundation models, which we will learn to implement in later chapters.

However, before we begin our practice project, we can trace its impact on the current landscape of GenAI. We follow the trajectory of early applications of the architecture (for example, **Bidirectional Encoded Representations from Transformers (BERT)**) through to the first GPT.

Evolving language models – the AR Transformer and its role in GenAI

In *Chapter 2*, we reviewed some of the generative paradigms that apply a transformer-based approach. Here, we trace the evolution of Transformers more closely, outlining some of the most impactful transformer-based language models from the initial transformer in 2017 to more recent state-of-the-art models that demonstrate the scalability, versatility, and societal considerations involved in this fast-moving domain of AI (as illustrated in *Figure 3.3*):

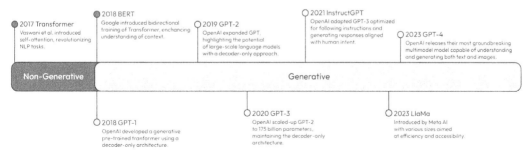

Figure 3.3: From the original transformer to GPT-4

- **2017 – Transformer**: The transformer model, introduced by Vaswani et al., was a paradigm shift in NLP, featuring self-attention layers that could process entire sequences of data in parallel. This architecture enabled the model to evaluate the importance of each word in a sentence relative to all other words, thereby enhancing the model's ability to capture the context.

- **2018 – BERT**: Google's BERT model innovated on the transformer architecture by utilizing a bidirectional context in its encoder layers during pre-training. It was one of the first models to understand the context of a word based on its entire sentence, both left and right, significantly improving performance on a wide range of NLP tasks, especially those requiring a deep understanding of context.

- **2018 – GPT-1**: OpenAI's GPT-1 model was a milestone in NLP, adopting a generative pre-trained approach with a transformer's decoder-only model. It was pre-trained on a diverse corpus of text data and fine-tuned for various tasks, using a unidirectional approach that generated text sequentially from left to right, which was particularly suited for generative text applications.

- **2019 – GPT-2**: GPT-2 built upon the foundation laid by GPT-1, maintaining its decoder-only architecture but significantly expanding its scale in terms of dataset and model size. This allowed GPT-2 to generate text that was more coherent and contextually relevant across a broader range of topics, demonstrating the power of scaling up transformer models.

- **2020 – GPT-3**: OpenAI's GPT-3 pushed the boundaries of scale in transformer models to 175 billion parameters, enabling a wide range of tasks to be performed with minimal input, often with **zero-shot learning (ZSL)** or **few-shot learning (FSL)**. This showed that Transformers could generalize across tasks and data types, often without the need for extensive task-specific data or fine-tuning.

- **2021 – InstructGPT**: An optimized variant of GPT-3, InstructGPT was fine-tuned specifically to follow user instructions and generate aligned responses, incorporating feedback loops that emphasized safety and relevance in its outputs. This represented a focus on creating AI models that could more accurately interpret and respond to human prompts.

- **2023 – GPT-4**: GPT-4 was an evolution of OpenAI's transformer models into the multimodal space, capable of understanding and generating content based on both text and images. This model aimed to produce safer and more contextually nuanced responses, showcasing a significant advancement in the model's ability to handle complex tasks and generate creative content.

- **2023 – LLaMA 2**: Meta AI's LLaMA 2 was part of a suite of models that focused on efficiency and accessibility, allowing for high-performance language modeling while being more resource-efficient. This model was aimed at facilitating a broader range of research and application development within the AI community.

- **2023 – Claude 2**: Anthropic's Claude 2 was an advancement over Claude 1, increasing its token context window and improving its reasoning and memory capabilities. It aimed to align more closely with human values, offering responsible and nuanced generative capabilities for open-domain question-answering and other conversational AI applications, marking progress in ethical AI development.

The timeline presented highlights the remarkable progress in transformer-based language models over the past several years. What originated as an architecture that introduced the concept of self-attention has rapidly evolved into models with billions of parameters that can generate coherent text, answer questions, and perform a variety of intellectual tasks at high levels of performance. The increase in scale and accessibility of models such as GPT-4 has opened new possibilities for AI applications. At the same time, recent models have illustrated a focus on safety and ethics and providing more nuanced, helpful responses to users.

In the next section, we accomplish a rite of passage for practitioners with an interest in the NL field. We implement the key components of the original transformer architecture using Python to more fully understand the mechanics that started it all.

Implementing the original Transformer

The following code demonstrates how to implement a minimal transformer model for a Seq2Seq translation task, mainly translating English text to French. The code is structured into multiple sections, handling various aspects from data loading to model training and translation.

Data loading and preparation

Initially, the code loads a dataset and prepares it for training. The data is loaded from a CSV file, which is then split into English and French text. The text is limited to 100 characters for demonstration purposes to reduce training time. The CSV file includes a few thousand example data points and can be found in the book's GitHub repository (`https://github.com/PacktPublishing/Python-Generative-AI`) along with the complete code:

```
import pandas as pd
import numpy as np

# Load demo data
data = pd.read_csv("./Chapter_3/data/en-fr_mini.csv")

# Separate English and French lexicons
EN_TEXT = data.en.to_numpy().tolist()
FR_TEXT = data.fr.to_numpy().tolist()

# Arbitrarily cap at 100 characters for demonstration to avoid long
training times
def demo_limit(vocab, limit=100):
    return [i[:limit] for i in vocab]
```

```
EN_TEXT = demo_limit(EN_TEXT)
FR_TEXT = demo_limit(FR_TEXT)

# Establish the maximum length of a given sequence
MAX_LEN = 100
```

Tokenization

Next, a tokenizer is trained on the text data. The tokenizer is essential for converting text data into numerical data that can be fed into the model:

```
from tokenizers import Tokenizer
from tokenizers.models import WordPiece
from tokenizers.trainers import WordPieceTrainer
from tokenizers.pre_tokenizers import Whitespace

def train_tokenizer(texts):
    tokenizer = Tokenizer(WordPiece(unk_token="[UNK]"))
    tokenizer.pre_tokenizer = Whitespace()
    trainer = WordPieceTrainer(
        vocab_size=5000,
        special_tokens=["[PAD]", "[UNK]", "[CLS]", "[SEP]", "[MASK]",
            "<sos>", "<eos>"],
    )
    tokenizer.train_from_iterator(texts, trainer)
    return tokenizer

en_tokenizer = train_tokenizer(EN_TEXT)
fr_tokenizer = train_tokenizer(FR_TEXT)
```

Data tensorization

The text data is then tensorized, which involves converting the text data into tensor format. This step is crucial for preparing the data for training with PyTorch:

```
import torch
from torch.nn.utils.rnn import pad_sequence

def tensorize_data(text_data, tokenizer):
    numericalized_data = [
        torch.tensor(tokenizer.encode(text).ids) for text in text_data
    ]
    padded_data = pad_sequence(numericalized_data,
        batch_first=True)
```

```
        return padded_data

src_tensor = tensorize_data(EN_TEXT, en_tokenizer)
tgt_tensor = tensorize_data(FR_TEXT, fr_tokenizer)
```

Dataset creation

A custom dataset class is created to handle the data. This class is essential for loading the data in batches during training:

```
from torch.utils.data import Dataset, DataLoader

class TextDataset(Dataset):
    def __init__(self, src_data, tgt_data):
        self.src_data = src_data
        self.tgt_data = tgt_data

    def __len__(self):
        return len(self.src_data)

    def __getitem__(self, idx):
        return self.src_data[idx], self.tgt_data[idx]

dataset = TextDataset(src_tensor, tgt_tensor)
```

Embeddings layer

The embeddings layer maps each token to a continuous vector space. This layer is crucial for the model to understand and process the text data:

```
import torch.nn as nn

class Embeddings(nn.Module):
    def __init__(self, d_model, vocab_size):
        super(Embeddings, self).__init__()
        self.embed = nn.Embedding(vocab_size, d_model)

    def forward(self, x):
        return self.embed(x)
```

Positional encoding

The positional encoding layer adds positional information to the embeddings, which helps the model understand the order of tokens in the sequence:

```
import math

class PositionalEncoding(nn.Module):
    def __init__(self, d_model, dropout=0.1,
                    max_len=MAX_LEN
    ):
        super(PositionalEncoding, self).__init__()
        self.dropout = nn.Dropout(p=dropout)
        pe = torch.zeros(max_len, d_model)
        position = torch.arange(0.0, max_len).unsqueeze(1)
        div_term = torch.exp(
            torch.arange(0.0, d_model, 2) * - \
                (math.log(10000.0) / d_model)
        )
        pe[:, 0::2] = torch.sin(position * div_term)
        pe[:, 1::2] = torch.cos(position * div_term)
        pe = pe.unsqueeze(0)
        self.register_buffer("pe", pe)

    def forward(self, x):
        x = x + self.pe[:, : x.size(1)]
        return self.dropout(x)
```

Multi-head self-attention

The **multi-head self-attention** (**MHSA**) layer is a crucial part of the transformer architecture that allows the model to focus on different parts of the input sequence when producing an output sequence:

```
class MultiHeadSelfAttention(nn.Module):
    def __init__(self, d_model, nhead):
        super(MultiHeadSelfAttention, self).__init__()
        self.attention = nn.MultiheadAttention(d_model, nhead)

    def forward(self, x):
        return self.attention(x, x, x)
```

FFN

The FFN is a simple **fully connected NN (FCNN)** that operates independently on each position:

```
class FeedForward(nn.Module):
    def __init__(self, d_model, d_ff):
        super(FeedForward, self).__init__()
        self.linear1 = nn.Linear(d_model, d_ff)
        self.dropout = nn.Dropout(0.1)
        self.linear2 = nn.Linear(d_ff, d_model)

    def forward(self, x):
        return self.linear2(self.dropout(torch.relu(self.linear1(x))))
```

Encoder layer

The encoder layer consists of an MHSA mechanism and a simple FFNN. This structure is repeated in a stack to form the complete encoder:

```
class EncoderLayer(nn.Module):
    def __init__(self, d_model, nhead, d_ff):
        super(EncoderLayer, self).__init__()
        self.self_attn = MultiHeadSelfAttention(d_model, nhead)
        self.feed_forward = FeedForward(d_model, d_ff)
        self.norm1 = nn.LayerNorm(d_model)
        self.norm2 = nn.LayerNorm(d_model)
        self.dropout = nn.Dropout(0.1)

    def forward(self, x):
        x = x.transpose(0, 1)
        attn_output, _ = self.self_attn(x)
        x = x + self.dropout(attn_output)
        x = self.norm1(x)
        ff_output = self.feed_forward(x)
        x = x + self.dropout(ff_output)
        return self.norm2(x).transpose(0, 1)
```

Encoder

The encoder is a stack of identical layers with an MHSA mechanism and an FFN:

```
class Encoder(nn.Module):
    def __init__(self, d_model, nhead, d_ff, num_layers, vocab_size):
        super(Encoder, self).__init__()
```

```
        self.embedding = Embeddings(d_model, vocab_size)
        self.pos_encoding = PositionalEncoding(d_model)
        self.encoder_layers = nn.ModuleList(
            [EncoderLayer(d_model, nhead, d_ff) for _ in range(
                num_layers)]
        )
        self.feed_forward = FeedForward(d_model, d_ff)

    def forward(self, x):
        x = self.embedding(x)
        x = self.pos_encoding(x)
        for layer in self.encoder_layers:
            x = layer(x)
        return x
```

Decoder layer

Similarly, the decoder layer consists of two MHA mechanisms—one self-attention and one cross-attention—followed by an FFN:

```
class DecoderLayer(nn.Module):
    def __init__(self, d_model, nhead, d_ff):
        super(DecoderLayer, self).__init__()
        self.self_attn = MultiHeadSelfAttention(d_model, nhead)
        self.cross_attn = nn.MultiheadAttention(d_model, nhead)
        self.feed_forward = FeedForward(d_model, d_ff)
        self.norm1 = nn.LayerNorm(d_model)
        self.norm2 = nn.LayerNorm(d_model)
        self.norm3 = nn.LayerNorm(d_model)
        self.dropout = nn.Dropout(0.1)

    def forward(self, x, memory):
        x = x.transpose(0, 1)
        memory = memory.transpose(0, 1)
        attn_output, _ = self.self_attn(x)
        x = x + self.dropout(attn_output)
        x = self.norm1(x)
        attn_output, _ = self.cross_attn(x, memory, memory)
        x = x + self.dropout(attn_output)
        x = self.norm2(x)
        ff_output = self.feed_forward(x)
        x = x + self.dropout(ff_output)
        return self.norm3(x).transpose(0, 1)
```

Decoder

The decoder is also a stack of identical layers. Each layer contains two MHA mechanisms and an FFN:

```python
class Decoder(nn.Module):
    def __init__(self, d_model, nhead, d_ff, num_layers, vocab_size):
        super(Decoder, self).__init__()
        self.embedding = Embeddings(d_model, vocab_size)
        self.pos_encoding = PositionalEncoding(d_model)
        self.decoder_layers = nn.ModuleList(
            [DecoderLayer(d_model, nhead, d_ff) for _ in range(
                num_layers)]
        )
        self.linear = nn.Linear(d_model, vocab_size)
        self.softmax = nn.Softmax(dim=2)

    def forward(self, x, memory):
        x = self.embedding(x)
        x = self.pos_encoding(x)
        for layer in self.decoder_layers:
            x = layer(x, memory)
        x = self.linear(x)
        return self.softmax(x)
```

This stacking layer pattern continues to build the transformer architecture. Each block has a specific role in processing the input data and generating output translations.

Complete transformer

The transformer model encapsulates the previously defined encoder and decoder structures. This is the primary class that will be used for training and translation tasks:

```python
class Transformer(nn.Module):
    def __init__(
        self,
        d_model,
        nhead,
        d_ff,
        num_encoder_layers,
        num_decoder_layers,
        src_vocab_size,
        tgt_vocab_size,
    ):
        super(Transformer, self).__init__()
```

```
        self.encoder = Encoder(d_model, nhead, d_ff, \
            num_encoder_layers, src_vocab_size)
        self.decoder = Decoder(d_model, nhead, d_ff, \
            num_decoder_layers, tgt_vocab_size)

    def forward(self, src, tgt):
        memory = self.encoder(src)
        output = self.decoder(tgt, memory)
        return output
```

Training function

The `train` function iterates through the epochs and batches, calculates the loss, and updates the model parameters:

```
def train(model, loss_fn, optimizer, NUM_EPOCHS=10):
    for epoch in range(NUM_EPOCHS):
        model.train()
        total_loss = 0
        for batch in batch_iterator:
            src, tgt = batch
            optimizer.zero_grad()
            output = model(src, tgt)
            loss = loss_fn(output.view(-1, TGT_VOCAB_SIZE),
                tgt.view(-1))
            loss.backward()
            optimizer.step()
            total_loss += loss.item()

        print(f"Epoch {epoch},
            Loss {total_loss / len(batch_iterator)}")
```

Translation function

The `translate` function uses the trained model to translate a source text into the target language. It generates a translation token by token and stops when an **end-of-sequence** (**EOS**) token is generated or when the maximum target length is reached:

```
def translate(model, src_text, src_tokenizer,
              tgt_tokenizer, max_target_length=50
):
    model.eval()

    src_tokens = src_tokenizer.encode(src_text).ids
```

```
src_tensor = torch.LongTensor(src_tokens).unsqueeze(0)

tgt_sos_idx = tgt_tokenizer.token_to_id("<sos>")
tgt_eos_idx = tgt_tokenizer.token_to_id("<eos>")

tgt_tensor = torch.LongTensor([tgt_sos_idx]).unsqueeze(0)

for i in range(max_target_length):
    with torch.no_grad():
        output = model(src_tensor, tgt_tensor)

    predicted_token_idx = output.argmax(dim=2)[0, -1].item()
    if predicted_token_idx == tgt_eos_idx:
        break
    tgt_tensor = torch.cat((tgt_tensor,
        torch.LongTensor([[predicted_token_idx]])),
        dim=1)

translated_token_ids = tgt_tensor[0, 1:].tolist()
translated_text = tgt_tokenizer.decode(translated_token_ids)

return translated_text
```

Main execution

In the main block of the script, hyperparameters are defined, the tokenizer and model are instantiated, and training and translation processes are initiated:

```
if __name__ == "__main__":
    NUM_ENCODER_LAYERS = 2
    NUM_DECODER_LAYERS = 2
    DROPOUT_RATE = 0.1
    EMBEDDING_DIM = 512
    NHEAD = 8
    FFN_HID_DIM = 2048
    BATCH_SIZE = 31
    LEARNING_RATE = 0.001

    en_tokenizer = train_tokenizer(EN_TEXT)
    fr_tokenizer = train_tokenizer(FR_TEXT)

    SRC_VOCAB_SIZE = len(en_tokenizer.get_vocab())
    TGT_VOCAB_SIZE = len(fr_tokenizer.get_vocab())
```

```
src_tensor = tensorize_data(EN_TEXT, en_tokenizer)
tgt_tensor = tensorize_data(FR_TEXT, fr_tokenizer)

dataset = TextDataset(src_tensor, tgt_tensor)

model = Transformer(
    EMBEDDING_DIM,
    NHEAD,
    FFN_HID_DIM,
    NUM_ENCODER_LAYERS,
    NUM_DECODER_LAYERS,
    SRC_VOCAB_SIZE,
    TGT_VOCAB_SIZE,
)
loss_fn = nn.CrossEntropyLoss()
optimizer = optim.Adam(model.parameters(), lr=LEARNING_RATE)

batch_iterator = DataLoader(
    dataset, batch_size=BATCH_SIZE,
    shuffle=True, drop_last=True
)

train(model, loss_fn, optimizer, NUM_EPOCHS=10)

src_text = "hello, how are you?"
translated_text = translate(
    model, src_text, en_tokenizer, fr_tokenizer)
print(translated_text)
```

This script orchestrates a machine translation task from loading data to training a transformer model and eventually translating text from English to French. Initially, it loads a dataset, processes the text, and establishes tokenizers to convert text to numerical data. Following this, it defines the architecture of a transformer model in PyTorch, detailing each component from the embeddings' self-attention mechanisms to the encoder and decoder stacks.

The script further organizes the data into batches, sets up a training loop, and defines a translation function. Training the model on the provided English and French sentences teaches it to map sequences from one language to another. Finally, it translates a sample sentence from English to French to demonstrate the model's capabilities.

Summary

The advent of the transformer significantly propelled the field of NLP forward, serving as the foundation for today's cutting-edge generative language models. This chapter delineated the progression of NLP that paved the way for this pivotal innovation. Initial statistical techniques such as count vectors and TF-IDF were adept at extracting rudimentary word patterns, yet they fell short in grasping semantic nuances.

Incorporating neural language models marked a stride toward more profound representations through word embeddings. Nevertheless, recurrent networks encountered hurdles in handling longer sequences. This inspired the emergence of CNNs, which introduced computational efficacy via parallelism, albeit at the expense of global contextual awareness.

The inception of attention mechanisms emerged as a cornerstone. In 2017, Vaswani et al. augmented these advancements, unveiling the transformer architecture. The hallmark self-attention mechanism of the transformer facilitates contextual modeling across extensive sequences in a parallelized manner. The layered encoder-decoder structure meticulously refines representations to discern relationships indispensable for endeavors such as translation.

The transformer, with its parallelizable and scalable self-attention design, set new benchmarks in performance. Its core tenets are the architectural bedrock for contemporary high-achieving generative language models such as GPT.

In the next chapter, we will discuss how to apply pre-trained generative models from prototype to production.

References

This reference section serves as a repository of sources referenced within this book; you can explore these resources to further enhance your understanding and knowledge of the subject matter:

- Bahdanau, D., Cho, K., and Bengio, Y. (2014). *Neural machine translation by jointly learning to align and translate*. arXiv preprint arXiv:1409.0473.

- Bengio, Y., Ducharme, R., and Vincent, P. (2003). *A neural probabilistic language model. The Journal of Machine Learning Research*, 3, 1137-1155.

- Dadgar, S. M. H., Araghi, M. S., and Farahani, M. M. (2016). *Improving text classification performance based on TFIDF and LSI index*. 2016 IEEE International Conference on Engineering & Technology (ICETECH).

- Elman, J. L. (1990). *Finding structure in time. Cognitive science*, 14(2), 179-211.

- Hochreiter, S., and Schmidhuber, J. (1997). *Long short-term memory. Neural computation*, 9(8), 1735-1780.

- Kim, Y. (2014). *Convolutional neural networks for sentence classification*. arXiv preprint arXiv:1408.5882.

- Mikolov, T., Sutskever, I., Chen, K., Corrado, G. S., and Dean, J. (2013). *Distributed representations of words and phrases and their compositionality. Advances in neural information processing systems*, 26.

- Pennington, J., Socher, R., and Manning, C. (2014). *GloVe: Global vectors for word representation. Proceedings of the 2014 conference on empirical methods in natural language processing (EMNLP)*, 1532-1543.

- Vaswani, A., Shazeer, N., Parmar, N., Uszkoreit, J., Jones, L., Gomez, A. N., ... and Polosukhin, I. (2017). *Attention is all you need. Advances in neural information processing systems*, 30.

4

Applying Pretrained Generative Models: From Prototype to Production

In the preceding chapters, we explored the fundamentals of generative AI, explored various generative models, such as **generative adversarial networks (GANs)**, diffusers, and transformers, and learned about the transformative impact of **natural language processing (NLP)**. As we transition into the practical aspects of applying generative AI, we should ground our exploration in a practical example. This approach will provide a concrete context, making the technical aspects more relatable and the learning experience more engaging.

We will introduce "StyleSprint," a clothing shop looking to enhance its online presence. One way to achieve this is by crafting unique and engaging product descriptions for its various products. However, manually creating captivating descriptions for a large inventory is challenging. This situation is prime opportunity for the application of generative AI. By leveraging a pretrained generative model, StyleSprint can automate the crafting of compelling product descriptions, saving considerable time and enriching the online shopping experience for its customers.

As we step into the practical application of a pretrained generative **large language models (LLM)**, the first order of business is to set up a Python environment conducive to prototyping with generative models. This setup is vital for transitioning the project from a prototype to a production-ready state, setting the stage for StyleSprint to realize its goal of automated content generation.

In *Chapters 2* and *3*, we used Google Colab for prototyping due to its ease of use and accessible GPU resources. It served as a great platform to test ideas quickly. However, as we shift our focus toward deploying our generative model in a real-world setting, it is essential to understand the transition from a prototyping environment such as Google Colab to a more robust, production-ready setup. This transition will ensure our solution is scalable, reliable, and well-optimized for handling real-world traffic. In this chapter, we will walk through the steps in setting up a production-ready Python environment, underscoring the crucial considerations for a smooth transition from prototype to production.

By the end of this chapter, we will understand the process of taking a generative application from a prototyping environment to a production-ready setup. We will define a reliable and repeatable strategy for evaluating, monitoring, and deploying models to production.

Prototyping environments

Jupyter notebooks provide an interactive computing environment to combine code execution, text, mathematics, plots, and rich media into a single document. They are ideal for prototyping and interactive development, making them a popular choice among data scientists, researchers, and engineers. Here is what they offer:

- **Kernel**: At the heart of a Jupyter notebook is a kernel, a computational engine that executes the code contained in the notebook. For Python, this is typically an IPython kernel. This kernel remains active and maintains the state of your notebook's computations while the notebook is open.

- **Interactive execution**: Code cells allow you to write and execute code interactively, inspecting the results and tweaking the code as necessary.

- **Dependency management**: You can install and manage libraries and dependencies directly within the notebook using `pip` or `conda` commands.

- **Visualization**: You can embed plots, graphs, and other visualizations to explore data and results interactively.

- **Documentation**: Combining Markdown cells with code cells allows for well-documented, self-contained notebooks that explain the code and its output.

A drawback to Jupyter notebooks is that they typically rely on the computational resources of your personal computer. Most personal laptops and desktops are not optimized or equipped to handle computationally intensive processes. Having adequate computational resources is crucial for managing the computational complexity of experimenting with an LLM. Fortunately, we can extend the capabilities of a Jupyter notebook with cloud-based platforms that offer computational accelerators such as **graphics processing units** (**GPUs**) and **tensor processing units** (**TPUs**). For example, Google Colab instantly enhances Jupyter notebooks, making them conducive to computationally intensive experimentation. Here are some of the key features of a cloud-based notebook environment such as Google Colab:

- **GPU/TPU access**: Provides free or affordable access to GPU and TPU resources for accelerated computation, which is crucial when working with demanding machine learning models

- **Collaboration**: Permits easy sharing and real-time collaboration, similar to Google Docs

- **Integration**: Allows for easy storage and access to notebooks and data

Let's consider our StyleSprint scenario. We will want to explore a few different models to generate product descriptions before deciding on one that best fits StyleSprint's goals. We can set up a minimal working prototype in Google Colab to compare models. Again, cloud-based platforms provide an optimal and accessible environment for initial testing, experimentation, and even some lightweight training of models. Here is how we might initially set up a generative model to start experimenting with automated product description generation for StyleSprint:

```
# In a Colab or Jupyter notebook
!pip install transformers
# Google Colab Jupyter notebook
from transformers import pipeline

# Initialize a text generation pipeline with a generative model, say
GPT-Neo
text_generator = pipeline(
    'text-generation', model='EleutherAI/gpt-neo-2.7B')

# Example prompt for product description generation
prompt = "This high-tech running shoe with advanced cushioning and
support"

# Generating the product description
generated_text = text_generator(prompt, max_length=100, do_
sample=True)

# Printing the generated product description
print(generated_text[0]['generated_text'])
```

Output:

```
This high-tech running shoe with advanced cushioning and support
combines the best of traditional running shoes and the latest
technologies.
```

In this simple setup, we're installing the `transformers` library, which offers a convenient interface to various pretrained models. We then initialize a text generation pipeline with an open source version of GPT-Neo, capable of generating coherent and contextually relevant text. This setup serves as a starting point for StyleSprint to experiment with generating creative product descriptions on a small scale.

Later in this chapter, we will expand our experiment to evaluate and compare multiple pretrained generative models to determine which best meets our needs. However, before advancing further in our experimentation and prototyping, it is crucial to strategically pause and project forward. This deliberate forethought allows us to consider the necessary steps for effectively transitioning our experiment into a production environment. By doing so, we ensure a comprehensive view of the project from end to end, to align with long-term operational goals.

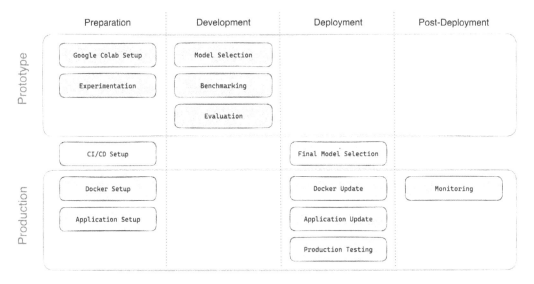

Figure 4.1: Moving from prototyping to production—the stages

Transitioning to production

As we plan for a production setup, we should first understand the intrinsic benefits and features of the prototyping environment we will want to carry forward to a production setting. Many of the features of prototyping environments such as Google Colab are deeply integrated and can easily go unnoticed, so it is important to dissect and catalog the features we will need in production. For example, the following features are inherent in Google Colab and will be critical in production:

- **Package management**: In Colab, installing necessary libraries is as straightforward as executing a cell with !pip install library _ name. In production, we will have to preinstall libraries or make sure we can install them as needed. We must also ensure that project-specific libraries do not interfere with other projects.

- **Dependency isolation**: Google Colab automatically facilitates isolated dependencies, ensuring package installations and updates do not interfere with other projects. In production, we may also want to deploy various projects using the same infrastructure. Dependency isolation will be critical to prevent one project's dependency updates from impacting other projects.

- **Interactive code execution**: The interactive execution of code cells helps in testing individual code snippets, visualizing results, and debugging in real time. This convenience is not necessary in production but could be helpful for quick debugging.

- **Resource accessibility**: With Colab, access to GPUs and TPUs is simplified, which is crucial for running computation-intensive tasks. For production, we will want to examine our dynamic computational needs and provision the appropriate infrastructure.

- **Data integration**: Colab offers simple connectivity to data sources for analysis and modeling. In production, we can either bootstrap our environment with data (i.e., deploy data directly into the environment) or ensure connectivity to remote data sources as needed.

- **Versioning and collaboration**: Tracking versions of your project code with Google Colab can easily be accomplished using notebooks. Additionally, Colab is preconfigured to interact with Git. Git is a distributed version control system that is widely used for tracking changes in source code during software development. In production, we will also want to integrate Git to manage our code and synchronize it with a remote code repository such as GitHub or Bitbucket. Remote versioning ensures that our production environment always reflects the latest changes and enables ongoing collaboration.

- **Error handling and debugging**: In Colab, we have direct access to the Python runtime and can typically see error messages and tracebacks in real time to help identify and resolve issues. We will want the same level of visibility in production via adequate logging of system errors. In total, we want to carry over the convenience and simplicity of our Google Colab prototyping environment but provide the robustness and scalability required for production. To do so, we will map each of the key characteristics we laid out to a corresponding production solution. These key features should ensure a smooth transition for deploying StyleSprint's generative model for automated product description generation.

Mapping features to production setup

To ensure we can seamlessly transition our prototyping environment to production, we can leverage Docker, a leading containerization tool. **Containerization** tools package applications with their dependencies for consistent performance across different systems. A containerized approach will help us replicate Google Colab's isolated, uniform environments, ensuring reliability and reducing potential compatibility issues in production. The table that follows describes how we can map each of the benefits of our prototyping environment to a production analog:

Feature	Environment	
	Prototyping	Production
Package management	Inherent through preinstalled package managers	Docker streamlines application deployment and consistency across environments including package managers.
Dependency isolation	Inherent through notebooks	Docker can also ensure projects are cleanly isolated.

Feature	Environment	
	Prototyping	**Production**
Interactive code execution	Inherent through notebooks	Docker helps to maintain versions of Python that provide interactive code execution by default. However, we may want to connect an **integrated development environment** (**IDE**) to our production environment to interact with code remotely as needed.
Resource accessibility	Inherent for cloud-based notebooks	GPU-enabled Docker containers enhance production by enabling structured GPU utilization, allowing scalable, efficient model performance.
Data integration	Not inherent, and requires code-based integration	Integrating Docker with a remote data source, such as AWS S3 or Google Cloud Storage, provides secure and scalable solutions for importing and exporting data.
Versioning and collaboration	Inherent through notebooks and preconfigured for Git	Integrating Docker with platforms such as GitHub or GitLab enables code collaboration and documentation.
Error handling and debugging	Inherent through direct interactive access to runtime	We can embed Python libraries such as `logging` or Loguru in Docker deployments for enhanced error tracking in production.

Table 4.1: Transitioning features from Colab to production via Docker

Having mapped out the features of our prototyping environment to corresponding tools and practices for a production setup, we are now better prepared to implement a generative model for StyleSprint in a production-ready environment. The transition entails setting up a stable, scalable, and reproducible Python environment, a crucial step for deploying our generative model to automate the generation of product descriptions in a real-world setting. As discussed, we can leverage Docker in tandem with GitHub and its **continuous integration/continuous deployment** (**CI/CD**) capabilities, providing a robust framework for this production deployment. A CI pipeline automates the integration of code changes from multiple contributors into a shared repository. We pair CI with CD to automate the deployment of our code to a production environment.

Setting up a production-ready environment

So far, we have discussed how to bridge the gap between prototyping and production environments. Cloud-based environments such as Google Colab provide a wealth of features that are not inherently available in production. Now that we have a better understanding of those characteristics, the next step is to implement a robust production setup to ensure that our application can handle real-world traffic, scale as needed, and remain stable over time.

The tools and practices in a production environment differ significantly from those in a prototyping environment. In production, scalability, reliability, resource management, and security become paramount, whereas, in a prototyping environment, the models are only relied upon by a few users for experimentation. In production, we could expect large-scale consumption from divisions throughout the organization. For example, in the StyleSprint scenario, there may be multiple departments or sub-brands hoping to automate their product descriptions.

In the early stages of our StyleSprint project, we can use free and open source tools such as Docker and GitHub for tasks such as containerization, version control, and CI. These tools are offered and managed by a community of users, giving us a cost-effective solution. As StyleSprint expands, we might consider upgrading to paid or enterprise editions that offer advanced features and professional support. For the moment, our focus is on leveraging the capabilities of the open source versions. Next, we will walk through the practical implementation of these tools step by step. By the end, we will be ready to deploy a production-ready **model-as-a-service** (**MaaS**) for automatic product descriptions.

Local development setup

We begin by making sure we can connect to a production environment remotely. We can leverage an IDE, which is software that enables us to easily organize code and remotely connect to the production environment.

Visual Studio Code

Begin by installing **Visual Studio Code** (**VS Code**), a free code editor by Microsoft. It is preferred for its integrated Git control, terminal, and marketplace for extensions that enhance its functionality. It provides a conducive environment for writing, testing, and debugging code.

Project initialization

Next, we set up a structured project directory to keep the code modular and organized. We will also initialize our working directory with Git, which enables us to synchronize code with a remote repository. As mentioned, we leverage Git to keep track of code changes and collaborate with others more seamlessly. Using the terminal window in Visual Studio, we can initialize the project using three simple commands. We use `mkdir` to create or "make" a directory. We use the `cd` command to change directories. Finally, we use `git init` to initialize our project with Git. Keep in mind that this assumes Git is installed. Instructions to install Git are made available on its website (`https://git-scm.com/`).

```
mkdir StyleSprint
cd StyleSprint
git init
```

Docker setup

We'll now move on to setting up a Docker container. A Docker container is an isolated environment that encapsulates an application and its dependencies, ensuring consistent operation across different systems. For clarity, we can briefly describe the key aspects of Docker as follows:

- **Containers**: These are portable units comprising the application and its dependencies.

- **Host operating system's kernel**: When a Docker container is run on a host machine, it utilizes the kernel of the host's operating system and resources to operate, but it does so in a way that is isolated from both the host system and other containers.

- **Dockerfiles**: These are scripts used to create container images. They serve as a blueprint containing everything needed to run the application. This isolation and packaging method prevents application conflicts and promotes efficient resource use, streamlining development and deployment.

A containerized approach will help ensure consistency and portability. For example, assume StyleSprint finds a cloud-based hosting provider that is more cost-effective; moving to the new provider is as simple as migrating a few configuration files.

We can install Docker from the official website. Docker provides easy-to-follow installation guides including support for various programming languages.

Once Docker is installed, we can create a Dockerfile in the project directory to specify the environment setup. For GPU support, we will want to start from an NVIDIA CUDA base image. Docker, like many other virtualized systems, operates using a concept called **images**. Images are a snapshot of a preconfigured environment that can be used as a starting point for a new project. In our case, we will want to start with a snapshot that integrates GPU support using the CUDA library, which is a parallel processing library provided by NVIDIA. This library will enable the virtualized environment (or container) to leverage any GPUs installed on the host machine. Leveraging GPUs will accelerate model inferencing.

Now we can go ahead and create a Dockerfile with the specifications for our application:

```
# Use an official NVIDIA CUDA runtime as a base image
FROM nvidia/cuda:11.0-base

# Set the working directory in the container to /app
WORKDIR /app

# Copy the current directory contents into the container at /app
COPY . /app

# Install any needed packages specified in requirements.txt
RUN pip install --no-cache-dir -r requirements.txt
```

```
# Make port 80 available to the world outside this container
EXPOSE 80


# Run app.py when the container launches
CMD ["uvicorn", "app:app", "--host", "0.0.0.0", "--port", "80"]
```

This Dockerfile serves as a blueprint that Docker follows to build our container. We initiate the process from an official NVIDIA CUDA base image to ensure GPU support. The working directory in the container is set to /app, where we then copy the contents of our project. Following that, we install the necessary packages listed in the `requirements.txt` file. Port 80 is exposed for external access to our application. Lastly, we specify the command to launch our application, which is running `app.py` using the Python interpreter. This setup encapsulates all the necessary components, including GPU support, to ensure our generative model operates efficiently in a production-like environment.

Requirements file

We also need a method for keeping track of our Python-specific dependencies. The container will include Python but will not have any indication as to what requirements our Python application has. We can specify those dependencies explicitly by defining a `requirements.txt` file in our project directory to list all the necessary Python packages:

```
fastapi==0.65.2
torch==1.9.0
transformers==4.9.2
uvicorn==0.14.0
```

Application code

Now we can create an `app.py` file for our application code. This is where we will write the code for our generative model, leveraging libraries such as PyTorch and Transformers. To expose our model as a service, we will use *FastAPI*, a modern, high-performance framework for building web APIs. A web API is a protocol that enables different software applications to communicate and exchange data over the internet, allowing them to use each other's functions and services.

The following snippet creates a minimal API that will serve the model responses whenever another application or software requests the `/generate/` endpoint. This will enable StyleSprint to host its model as a web service. This means that other applications (e.g., mobile apps, batch processes) can access the model using a simple URL. We can also add exception handling to provide an informative error message should the model produce any kind of error:

```
from fastapi import FastAPI, HTTPException
from pydantic import BaseModel
from transformers import pipeline
```

```
# Load the pre-trained model
generator = pipeline('text-generation',
    model='EleutherAI/gpt-neo-2.7B')

# Create the FastAPI app
app = FastAPI()

# Define the request body
class GenerationInput(BaseModel):
prompt: str

# Define the endpoint
@app.post("/generate")
def generate_text(input: GenerationInput):
try:
    # Generate text based on the input prompt
    generated_text = generator(input.prompt, max_length=150)
    return {"generated_text": generated_text}
except:
    raise HTTPException(status_code=500,
        detail="Model failed to generate text")
```

Now that we have a Docker setup, the next step is to deploy the application to the host server. We can streamline this process with a CI/CD pipeline. The goal is to fully automate all deployment steps, including a suite of tests to ensure that any code changes do not introduce any errors. We then leverage GitHub Actions to create a workflow that is directly integrated with a code repository.

Creating a code repository

Before we can leverage the automation capabilities of GitHub, we will need a repository. Creating a GitHub repository is straightforward, following these steps:

1. **Sign up/log in to GitHub**: If you don't have a GitHub account, sign up at github.com. If you already have an account, just log in.

2. **Go to the repository creation page**: Click the + icon in the top-right corner of the GitHub home page and select **New repository**.

3. **Fill in the repository details**:

    * **Repository Name**: Choose a name for your repository

    * **Description** (optional): Add a brief description of your repository

    * **Visibility**: Select either **Public** (anyone can see this repository) or **Private** (only you and the collaborators you invite can see it)

4. **Initialize the repository with a README** (optional):

   - Check **Initialize this repository with a README** if you want to add a simple text file that can be updated later to provide instructions for collaborators.

   - We can also add a .gitignore file or choose a license. A gitignore file allows us to add paths or file types that should not be uploaded to the repository. For example, Python creates temporary files that are not critical to the application. Adding `` `_ _ pycache _ _ /` `` to the gitignore file will automatically ignore all contents of that directory.

5. **Create repository**: Click the **Create repository** button.

With our repository setup complete, we can move on to defining our CI/CD pipeline to automate our deployments.

CI/CD setup

To create a pipeline, we will need a configuration file that outlines the stages of deployment and instructs the automation server to build and deploy our Docker container. Let's look at the steps:

1. In our GitHub repository, we can create a new file in the .github/workflows directory named ci-cd.yml. GitHub will automatically find any files in this directory to trigger deployments.

2. Open ci-cd.yml and define the following workflow:

```
name: CI/CD Pipeline

on:
  push:
    branches:
      - main

jobs:
  build-and-test:
    runs-on: ubuntu-latest

    steps:
      - name: Checkout code
        uses: actions/checkout@v4

      - name: Build Docker image
# assumes the Dockerfile is in the root (.)
        run: docker build -t stylesprint .

      - name: Run tests
# assumes a set of unit tests were defined
```

```
        run: docker run stylesprint python -m unittest discover

deploy:
  needs: build-and-test
  runs-on: ubuntu-latest

  steps:
    - name: Checkout code
      uses: actions/checkout@v4

    - name: Login to DockerHub
      run: echo ${{ secrets.DOCKER_PASSWORD }} | docker login -u
${{ secrets.DOCKER_USERNAME }} --password-stdin

    - name: Push Docker image
      run: |
        docker tag stylesprint:latest ${{ secrets.DOCKER_
USERNAME }}/stylesprint:latest
        docker push ${{ secrets.DOCKER_USERNAME }}/
stylesprint:latest
```

In this setup, our workflow consists of two primary jobs: build-and-test and deploy. The build-and-test job is responsible for checking out the code from the repository, building the Docker image, and executing any tests. On the other hand, the deploy job, which relies on completing build-and-test, handles *DockerHub* login and pushes the Docker image there. DockerHub, similar to GitHub, is a repository specifically for Docker images.

For authenticating with DockerHub, it is advised to securely store your DockerHub credentials in your GitHub repository. This can be done by navigating to your repository on GitHub, clicking on **Settings**, then **Secrets**, and adding DOCKER_USERNAME and DOCKER_PASSWORD as new repository secrets.

Notice that we did not have to perform any additional steps to execute the pipeline. The workflow is designed to trigger automatically upon a push (or upload) to the main branch. Recall that the entire process relies on the Git pattern where new changes are registered through a commit or check-in of code and a push or upload of code changes. Whenever changes are pushed, we can directly observe the entire pipeline in action within the **Actions** tab of the GitHub repository.

We have now walked through all of the steps necessary to deploy our model to production. With all of this critical setup behind us, we can now return to choosing the best model for our project. The goal is to find a model that can effectively generate captivating product descriptions for StyleSprint. However, the variety of generative models available requires a thoughtful choice based on our project's needs and constraints.

Moreover, we want to choose the right evaluation metrics and discuss other considerations that will guide us in making an informed decision for our project. This exploration will equip us with the knowledge needed to select a model that not only performs well but also aligns with our project objectives and the technical infrastructure we have established.

Model selection – choosing the right pretrained generative model

Having established a minimal production environment in the previous section, we now focus on a pivotal aspect of our project – selecting the right generative model for generating engaging product descriptions. The choice of model is crucial as it significantly influences the effectiveness and efficiency of our solution. The objective is to automate the generation of compelling and accurate product descriptions for StyleSprint's diverse range of retail products. By doing so, we aim to enrich the online shopping experience for customers while alleviating the manual workload of crafting unique product descriptions.

Our objective is to select a generative model that can adeptly handle nuanced and sophisticated text generation to significantly expedite the process of creating unique, engaging product descriptions, saving time and resources for StyleSprint.

In selecting our model, it is important to thoroughly evaluate various factors influencing its performance and suitability for the project.

Meeting project objectives

Before we can select and apply evaluation methods to our model selection process, we should first make sure we understand the project objectives. This involves defining the business problem, identifying any technical constraints, identifying any risk associated with the model, including interpretation of model outcomes, and ascertaining considerations for any potential disparate treatment or bias:

- **Problem definition**: In our scenario, the goal is to create accurate and engaging descriptions for a wide range of retail clothing. As StyleSprint's product range may expand, the system should scale seamlessly to accommodate a larger inventory without significantly increasing operational costs. Performance expectations include compelling descriptions to attract potential customers, accuracy to avoid misrepresentation, and prompt generation to maintain an up-to-date online catalog. Additionally, StyleSprint may apply personalized content descriptions based on a user's shopping history. This implies that the model may have to provide product descriptions in near-real-time.

- **Technical constraints**: To maximize efficiency, there should not be any noticeable delay (latency) in responses from the model API. The system should be capable of real-time updates to the online catalog (as needed), and the hardware should support quick text generation without compromising quality while remaining cost-effective, especially as the product range expands.

- **Transparency and openness**: Generally, pretrained models from developers who disclose architectures and training data sources are preferred, as this level of transparency allows StyleSprint to have a clear understanding of any risks or legal implications associated with model use. Additionally, any usage restrictions imposed by using models provided as APIs, such as request or token limitations, should be understood as they could hinder scalability for a growing catalog.

- **Bias and fairness**: Identifying and mitigating biases in model outputs to ensure fair and neutral representations is crucial, especially given StyleSprint's diverse target audience. Ensuring that the generated descriptions are culturally sensitive is of paramount importance. Fair representation ensures that the descriptions accurately and fairly represent the products to all potential customers, irrespective of their individual characteristics or social backgrounds.

- **Suitability of pretraining**: The underlying pretraining of generative models plays a significant role in their ability to generate meaningful and relevant text. Investigating the domains and data on which the models were pretrained or fine-tuned is important. A model pretrained on a broad dataset may be versatile but could lack domain-specific nuances. For StyleSprint, a model that is fine-tuned on fashion-related data or that has the ability to be fine-tuned on such data would be ideal to ensure the generated descriptions are relevant and appealing.

- **Quantitative metrics**: Evaluating the quality of generated product descriptions for StyleSprint necessitates a combination of lexical and semantic metrics. Lexical overlap metrics measure the lexical similarity between generated and reference texts. Specifically, **Bilingual Evaluation Understudy (BLEU)** emphasizes n-gram precision, **Recall-Oriented Understudy for Gisting Evaluation (ROUGE)** focuses on n-gram recall, and **Metric for Evaluation of Translation with Explicit Ordering (METEOR)** aims for a more balanced evaluation by considering synonyms and stemming. For contextual and semantic evaluation, we use similarity metrics to assess the semantic coherence and relevance of the generated descriptions, often utilizing embeddings to represent text in a way that captures its meaning.

We can further refine our assessment of the alignment between generated descriptions and product images using models such as **Contrastive Language-Image Pretraining (CLIP)**. Recall that we used CLIP in *Chapter 2* to score the compatibility between captions and a synthesized image. In this case, we can apply CLIP to measure whether our generated descriptions accurately reflect the visual aspects of the products. Collectively, these evaluation techniques provide objective methods for assessing the performance of the generative model in creating effective product descriptions for StyleSprint:

- **Qualitative metrics**: We introduce qualitative evaluation to measure nuances such as the engaging and creative nature of descriptions. We also want to ensure we consider equity and inclusivity in the generated content, which is critical to avoid biases or language that could alienate or offend certain groups. Methods for engagement assessment could include customer surveys or A/B testing, a systematic method for testing two competing solutions. Additionally, having a diverse group reviewing the content for equity and inclusivity could provide valuable insights. These steps help StyleSprint create captivating, respectful, and inclusive product descriptions, fostering a welcoming environment for all customers.

- **Scalability**: The computational resources required to run a model and the model's ability to scale with increasing data are vital considerations. Models that demand extensive computational power may not be practical for real-time generation of product descriptions, especially as the product range expands. A balance between computational efficiency and output quality is essential to ensure cost-effectiveness and scalability for StyleSprint.

- **Customization and fine-tuning capabilities**: The ability to fine-tune or customize the model on domain-specific data is crucial for better aligning with brand-specific requirements. Exploring the availability and ease of fine-tuning can significantly impact the relevance and quality of generated descriptions, ensuring that they resonate well with the brand identity and product range of StyleSprint. In practice, some models are too large to fine-tune without considerable resources, even when efficient methods are applied. We will explore fine-tuning considerations in detail in the next chapter.

Now that we have carefully considered how we might align the model to the project's goals, we are almost ready to evaluate our initial model selection against a few others to ensure we make the right choice. However, before benchmarking, we should dedicate time to understanding one vital aspect of the model selection process: model size and computational complexity.

Model size and computational complexity

The size of a generative model is often described by the number of parameters it has. Parameters in a model are the internal variables that are fine-tuned during the training process based on the training data. In the context of neural networks used in generative models, parameters typically refer to the weights and biases adjusted through training to minimize the discrepancy between predicted outputs and actual targets.

Moreover, a model with more parameters can capture more complex patterns in the data, often leading to better performance on the task at hand. While larger models often perform better in terms of the quality of the generated text, there's a point of diminishing returns beyond which increasing model size yields marginal improvements. Moreover, the increased size comes with its own set of challenges:

- **Computational complexity**: Larger models require more computational power and memory, during both training and inference (the phase where the model is used to make predictions or generate new data based on the learned parameters). This can significantly increase the costs and the time required to train and use the model, making it less suitable for real-time applications or resource-constrained environments.

The number of parameters significantly impacts the computational complexity of a model. Each parameter in a model is a variable that must be stored in memory during computation, during both training and inference. Here are some specific considerations for computational requirements:

- **Memory and storage**: The total size of the model in memory is the product of the number of parameters and the size of each parameter (typically a 32-bit or 64-bit float). For instance, a model with 100 million parameters, each represented by a 32-bit float, would require approximately 400 MB of memory (100 million * 32 bits = 400 million bits = 400 MB). Now consider a larger model, say with 10 billion parameters; the memory requirement jumps to 40 GB (10 billion * 32 bits = 40 billion bits = 40 GB). This requirement is just for the parameters and does not account for other data and overheads the model needs for its operations.

- **Loading into memory**: When a model is used for inference, its parameters must be loaded into the RAM of the machine it's running on. For a large model with 10 billion parameters, you would need a machine with enough RAM to accommodate the entire model, along with additional memory for the operational overhead, the input data, and the generated output. Suppose the model is too large to fit in memory. In that case, it may need to be **sharded** or distributed across multiple machines or loaded in parts, which can significantly complicate the deployment and operation of the model and also increase the latency of generating outputs.

- **Specialized hardware requirements**: Larger models require specialized hardware, such as powerful GPUs or TPUs, which could increase the project costs. As discussed, models with a large number of parameters require powerful computational resources for both training and inference. Hardware accelerators such as GPUs and TPUs are often employed to meet these demands. These hardware accelerators are designed to handle the parallel computation capabilities needed for the matrix multiplications and other operations inherent in neural network computations, speeding up the processing significantly compared to traditional **central processing units (CPUs)**.

Cloud-based infrastructure can alleviate the complexity of setup but often has usage-based pricing. Understanding infrastructure costs on a granular level is vital to ensuring that StyleSprint stays within its budget.

- **Latency**: We've briefly discussed latency, but it is important to reiterate that larger models typically have higher latency, which could be a problem for applications that require real-time responses. In our case, we can process the descriptions as batches asynchronously. However, StyleSprint may have projects that require fast turnarounds, requiring batches to be completed in hours and not days.

In the case of StyleSprint, the trade-off between model performance and size must be carefully evaluated to ensure the final model meets the project's performance requirements while staying within budget and hardware constraints. StyleSprint was hoping to have near-real-time responses to provide personalized descriptions, which typically translates to a smaller model with less computational complexity. However, it was also important that the model remains highly accurate and aligns with branding standards for tone and voice, which may require a larger model trained or fine-tuned on a larger dataset. In practice, we can evaluate the performance of models relative to size and complexity through benchmarking.

Benchmarking

Benchmarking is a systematic process used to evaluate the performance of different generative models against predefined criteria. This process involves comparing the models on various metrics to understand their strengths, weaknesses, and suitability for the project. It is an empirical method (based on observation) to obtain data on how the models perform under similar conditions, providing insights that can inform the decision-making process for model selection.

In the StyleSprint scenario, benchmarking can be an invaluable exercise to navigate the trade-offs between model size, computational complexity, and the accuracy and creativity of generated descriptions.

For our benchmarking exercise, we can return to our Google Colab prototyping environment to quickly load various generative models and run them through tests designed to evaluate their performance based on the considerations outlined in the previous sections, such as computational efficiency and text generation quality. Once we have completed our evaluation and comparison, we can make a few simple changes to our production application code and it will automatically redeploy. Benchmarking will be instrumental in measuring the quality of the descriptions relative to the model size and complexity. Recall that we will measure quality and overall model performance along several dimensions, including lexical and semantic similarity to a "gold standard" of human-written descriptions, and a qualitative assessment performed by a diverse group of reviewers.

The next step is to revisit and adapt our original prototyping code to include a few challenger models and apply evaluation metrics.

Updating the prototyping environment

For our evaluation steps, there are a few key changes to our original experimentation setup in Google Colab. First, we will want to make sure we leverage performance acceleration. Google Colab offers acceleration via GPU or TPU environments. For this experiment, we will leverage GPU. We will also want to transition from the Transformers library to a slightly more versatile library such as Langchain, which allows us to test both open source models such as GPT-Neo and commercial models such as GPT-3.5.

GPU configuration

Ensure you have a GPU enabled for better performance. Returning to Google Colab, we can follow these steps to enable GPU acceleration:

1. Click on **Runtime** in the top menu (see *Figure 4.2*):

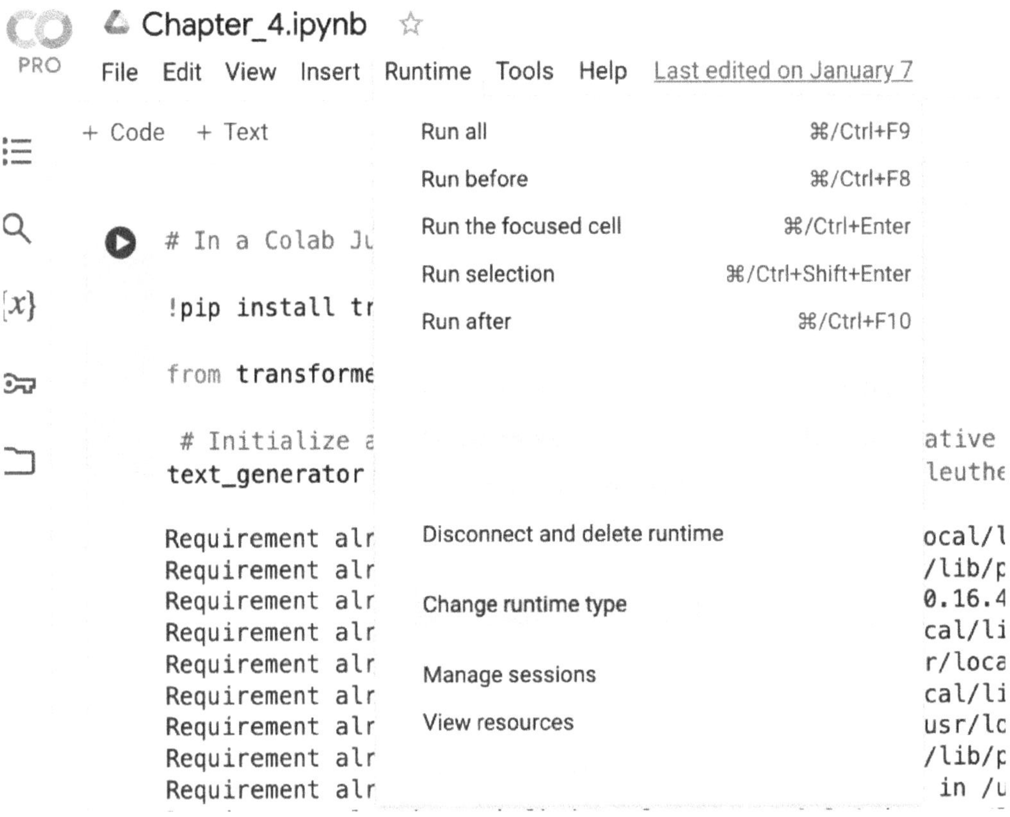

Figure 4.2: Runtime drop-down menu

2. Select **Change runtime type** from the drop-down menu, as shown in the preceding screenshot.

3. In the pop-up window, select **GPU** from the **Hardware accelerator** drop-down menu (see *Figure 4.3*):

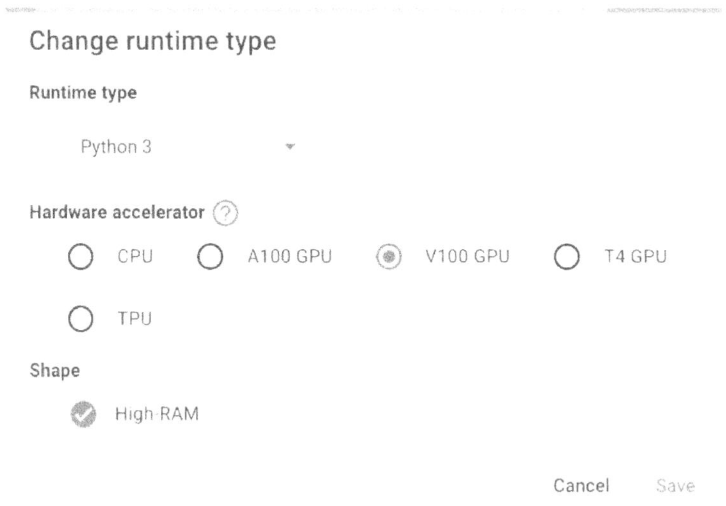

Figure 4.3: Select GPU and click on Save

4. Click on **Save**.

Now your notebook is set up to use a GPU to significantly speed up the computations needed for the benchmarking process. You can verify the GPU availability using the following code snippet:

```
# Verify GPU is available
import torch
torch.cuda.is _ available()
```

This code snippet will return `True` if a GPU is available and `False` otherwise. This setup ensures that you have the necessary computational resources to benchmark various generative models. The utilization of a GPU will be crucial when it comes to handling large models and extensive computations.

Loading pretrained models with LangChain

In our first simple experiment, we relied on the Transformers library to load an open source version of GPT. However, for our benchmarking exercise, we want to evaluate the retail version of GPT-3 alongside open source models. We can leverage LangChain, a versatile library that provides a streamlined interface, to access both open source models from providers such as Hugging Face and closed source models such as OpenAI's GPT-3.5. LangChain offers a unified API that simplifies benchmarking and comparison through standardization. Here are the steps to do it:

1. **Install necessary libraries**: We begin by installing the required libraries in our Colab environment. LangChain simplifies the interaction with models hosted on OpenAI and Hugging Face.

    ```
    !pip -q install openai langchain huggingface _ hub
    ```

2. **Set up credentials**: We obtain the credentials from OpenAI for accessing GPT-3, GPT-4, or whichever closed source model we select. We also provide credentials for the Hugging Face Hub, which hosts over 350,000 open source models. We must store these credentials securely to prevent any unauthorized access, especially in the case where model usage has an associated cost.

    ```
    import os

    os.environ['OPENAI _ API _ KEY'] = 'your _ openai _ api _ key _ here'
    os.environ['HUGGINGFACEHUB _ API _ TOKEN'] =
        'your _ huggingface _ token _ here'
    ```

3. **Load models**: With LangChain, we can quickly load models and generate responses. The following example demonstrates how to load GPT-3 and GPT-Neo from Hugging Face:

    ```
    !pip install openai langchain[llms] huggingface _ hub

    from langchain.llms import OpenAI, HuggingFaceHub

    # Loading GPT-3
    llm _ gpt3 = OpenAI(model _ name='text-davinci-003',
                        temperature=0.9,
                        max _ tokens = 256)

    # Loading Neo from Hugging Face
    llm _ neo = HuggingFaceHub(repo _ id=' EleutherAI/gpt-neo-2.7B',
                              model _ kwargs={"temperature":0.9}
    )
    ```

Notice that we have loaded two models that are significantly different in size. As the model signature suggests, GPT-Neo was trained on 2.7 billion parameters. Meanwhile, according to information available from OpenAI, Davinci was trained on 175 billion parameters. As discussed, a model that is significantly larger is expected to have captured much more complex patterns and will likely outperform a smaller model. However, these very large models are typically hosted by major providers and have higher usage costs. We will revisit cost considerations later. For now, we can continue to the next step, which is to prepare our testing data. Our test data should provide a baseline for model performance that will inform the cost versus performance trade-off.

Setting up testing data

In this context, testing data should comprise product attributes from the StyleSprint website (e.g., available colors, sizes, materials, etc.) and existing product descriptions written by the StyleSprint team. The human-written descriptions serve as the "ground truth," or the standard against which to compare the models' generated descriptions.

We can gather product data from existing datasets by scraping data from e-commerce websites or using a pre-collected dataset from StyleSprint's database. We should also ensure a varied collection of products to test a model's capability across different categories and styles. The process of dividing data into distinct groups or segments based on shared characteristics is typically referred to as segmentation. Understanding a model's behavior across segments should give us an indication of how well it can perform across the entire family of products. For the purposes of this example, product data is made available in the GitHub companion to this book (`https://github.com/PacktPublishing/Generative-AI-Foundations-in-Python`).

Let's see how we can extract relevant information for further processing:

```python
import pandas as pd

# Assume `product_data.csv` is a CSV file with product data
# The CSV file has two columns: 'product_image' and
# 'product_description'

# Load the product data
product_data = pd.read_csv('product_data.csv')

# Split the data into testing and reference sets
test_data = product_data.sample(frac=0.2, random_state=42)
reference_data = product_data.drop(test_data.index)

# Checkpoint the testing and reference data
test_data.to_csv('test_data.csv', index=False)
reference_data.to_csv('reference_data.csv', index=False)
# Extract reference descriptions and image file paths
reference_descriptions = /
    reference_data['product_description'].tolist()
product_images = reference_data['product_image'].tolist()
```

We must also format the product data in a way that makes it ready to be input into the models for description generation. This could be just the product title or a combination of product attributes:

```python
# Assume `product_metadata` is a column in the data that contains the
collective information about the product including the title of the
product and attributes.
# Format the input data for the models
model_input_data = reference_data['product_metadata].tolist()
reference_descriptions = \
    reference_data['product_description'].tolist()
```

Finally, we will ask the model to generate a batch of product descriptions using each model.

```
from langchain import LLMChain, PromptTemplate
from tqdm.auto import tqdm

template = """
Write a creative product description for the following product:
{product _ metadata}
"""

PROMPT = PromptTemplate(template=template,
    input _ variables=["product _ metadata"])

def generate _ descriptions(
    llm: object,
    prompt: PromptTemplate = PROMPT
) -> list:
    # Initialize the LLM chain
    llm _ chain = LLMChain(prompt=prompt, llm=llm)
    descriptions = []
    for i in tqdm(range(len(model _ input _ data))):
        description = llm _ chain.run(model _ input _ data[i])
        descriptions.append(description)
    return descriptions

gpt3 _ descriptions = generate _ descriptions(llm _ gpt3)
gptneo _ descriptions = generate _ descriptions(llm _ neo)
```

Now, with the testing data set up, we have a structured dataset of product information, reference descriptions, and images ready for use in the evaluation steps.

Quantitative metrics evaluation

Now that we have leveraged Langchain to load multiple models and prepared testing data, we are ready to begin applying evaluation metrics. These metrics capture accuracy and alignment with product images and will help us assess how well the models generate product descriptions compared to humans. As discussed, we focused on two categories of metrics, lexical and semantic similarity, which provide a measure of how many of the same words were used and how much semantic information is common to both the human and AI-generated product descriptions.

In the following code block, we apply BLEU, ROUGE, and METEOR to evaluate the lexical similarity between the generated text and the reference text. Each of these has a reference-based assumption.

This means that each metric assumes we are comparing against a human reference. We have already set aside our reference descriptions (or gold standard) for a diverse set of products to compare side-by-side with the generated descriptions.

```
!pip install rouge sumeval nltk

# nltk requires an additional package
import nltk
nltk.download('wordnet')

 from nltk.translate.bleu_score import sentence_bleu

from rouge import Rouge
from sumeval.metrics.rouge import RougeCalculator
from nltk.translate.meteor_score import meteor_score

def evaluate(
    reference_descriptions: list,
    generated_descriptions: list
) -> tuple:
    # Calculating BLEU score
    bleu_scores = [
        sentence_bleu([ref], gen)
        for ref, gen in zip(reference_descriptions, generated_
descriptions)
    ]
    average_bleu = sum(bleu_scores) / len(bleu_scores)

    # Calculating ROUGE score
    rouge = RougeCalculator()
    rouge_scores = [rouge.rouge_n(gen, ref, 2) for ref,
        gen in zip(reference_descriptions,
        generated_descriptions)]
    average_rouge = sum(rouge_scores) / len(rouge_scores)

    # Calculating METEOR score
    meteor_scores = [ meteor_score([ref.split() ],
        gen.split()) for ref,
        gen in zip(reference_descriptions,
        generated_descriptions)]
    average_meteor = sum(meteor_scores) / len(meteor_scores)

    return average_bleu, average_rouge, average_meteor
```

```
average_bleu_gpt3, average_rouge_gpt3, average_meteor_gpt3 = \
    evaluate(reference_descriptions, gpt3_descriptions)
print(average_bleu_gpt3, average_rouge_gpt3, average_meteor_gpt3)

average_bleu_neo, average_rouge_neo, average_meteor_neo = \
    evaluate(reference_descriptions, gptneo_descriptions)
print(average_bleu_neo, average_rouge_neo, average_meteor_neo)
```

We can evaluate the semantic coherence and relevance of the generated descriptions using sentence embeddings:

```
!pip install sentence-transformers
from sentence_transformers import SentenceTransformer, util

model = SentenceTransformer('paraphrase-MiniLM-L6-v2')

def cosine_similarity(reference_descriptions, generated_
descriptions):

    # Calculating cosine similarity for generated descriptions

    cosine_scores = [util.pytorch_cos_sim(
        model.encode(ref), model.encode(gen)) for ref,
        gen in zip(reference_descriptions,
        generated_descriptions)]
    average_cosine = sum(cosine_scores) / len(cosine_scores)

    return average_cosine

average_cosine_gpt3 = cosine_similarity(
    reference_descriptions, gpt3_descriptions)
print(average_cosine_gpt3)

average_cosine_neo = cosine_similarity(
    reference_descriptions, gptneo_descriptions)
print(average_cosine_neo)
```

Alignment with CLIP

We again leverage the CLIP model to evaluate the alignment between generated product descriptions and corresponding images, similar to our approach in *Chapter 2*. The CLIP model, adept at correlating visual and textual content, scores the congruence between each product image and its associated generated and reference descriptions. The reference description serves as a human baseline for accuracy. These scores provide a quantitative measure of our generative model's effectiveness at producing descriptions that correspond well to the product image. The following is a snippet from a component that processes the generated descriptions combined with corresponding images to generate a CLIP score. The full component code (including image pre-processing) is available in the chapter 4 folder of this book's GitHub repository at https://github.com/PacktPublishing/Generative-AI-Foundations-in-Python).

```python
clip_model = "openai/clip-vit-base-patch32"
def clip_scores(images, descriptions,
                model=clip_model,
                processor=clip_processor
):
    scores = []

    # Process all images and descriptions together
    inputs = process_inputs(processor, descriptions, images)

    # Get model outputs
    outputs = model(**inputs)
    logits_per_image = outputs.logits_per_image # Image-to-text
logits

    # Diagonal of the matrix gives the scores for each image-
description pair
    for i in range(logits_per_image.size(0)):
        score = logits_per_image[i, i].item()
    scores.append(score)

    return scores

reference_images = [
    load_image_from_path(image_path)
    for image_path in reference_data.product_image_path
]

gpt3_generated_scores = clip_scores(
    reference_images, gpt3_descriptions
)

reference_scores = clip_scores(
    reference_images, reference_descriptions
)

# Compare the scores
for i, (gen_score, ref_score) in enumerate(
```

```
    zip(gpt3_generated_scores, reference_scores)
):
    print(f"Image {i}: Generated Score = {gen_score:.2f},
        Reference Score = {ref_score:.2f}")
```

In evaluating product descriptions using the CLIP model, the alignment scores generated for each image-description pair are computed relative to other descriptions in the batch. Essentially, CLIP assesses how well a specific description (either generated or reference) aligns with a given image compared to other descriptions within the same batch. For example, a score of 33.79 indicates that the description aligns with the image 33.79% better than the other descriptions in the batch align with that image. In comparing against the reference, we expect that the scores based on the generated descriptions should align closely with the scores based on the reference descriptions.

Now that we have calculated lexical and semantic similarity to the reference scores, and alignment between images and generated descriptions relative to reference descriptions, we can evaluate our models holistically and interpret the outcome of our quantitative evaluation.

Interpreting outcomes

We begin with lexical similarity, which gives us an indication of similarity in phrasing and keywords between the reference and generated descriptions:

	BLEU	**ROUGE**	**METEOR**
GPT-3.5	0.147	0.094	0.261
GPT-Neo	0.132	0.05	0.059

Table 4.2: Lexical similarity

In evaluating text generated by GPT-3.5 and GPT-Neo models, we use several lexical similarity metrics: BLEU, ROUGE, and METEOR. BLEU scores, which assess the precision of matching phrases, show GPT-3.5 (0.147) slightly outperforming GPT-Neo (0.132). ROUGE scores, focusing on the recall of content, indicate that GPT-3.5 (0.094) better captures reference content than GPT-Neo (0.05). METEOR scores, combining both precision and recall with synonym matching, reveal a significant lead for GPT-3.5 (0.261) over GPT-Neo (0.059). Overall, these metrics suggest that GPT-3.5's generated text aligns more closely with reference standards, both in word choice and content coverage, compared to that of GPT-Neo.

Next, we evaluate semantic similarity, which measures how closely the meanings of the generated text align with the reference text. This assessment goes beyond mere word-to-word matching and considers the context and overall intent of the sentences. Semantic similarity evaluates the extent to which the generated text captures the nuances, concepts, and themes present in the reference text, providing insight into the model's ability to understand and replicate deeper semantic meanings:

Model	Mean cosine similarity
GPT-3.5	0.8192
GPT-Neo	0.2289

Table 4.3: Semantic similarity

The mean cosine similarity scores reveal a stark contrast between the two models' performance in semantic similarity. GPT-3.5 shows a high degree of semantic alignment with the reference text. GPT-Neo's significantly lower score suggests a relatively poor performance, indicating that the generated descriptions were fundamentally dissimilar to descriptions written by humans.

Finally, we review the CLIP scores, which tell us how well the generated descriptions align visually with the corresponding images. These scores, derived from a model trained to understand and correlate visual and textual data, provide a measure of the relevance and accuracy of the text in representing the visual content. High CLIP scores indicate a strong correlation between the text and the image, suggesting that the generated descriptions are not only textually coherent but also contextually appropriate and visually descriptive:

Model	Mean CLIP	Reference delta
GPT-3.5	26.195	2.815
GPT-Neo	22.647	6.363

Table 4.4: Comparative CLIP score analysis for GPT-3.5 and GPT-Neo models

We calculated the CLIP scores from the reference descriptions, which represent the average alignment score between a set of benchmark descriptions and the corresponding images. We then calculated CLIP scores for each model and analyzed the delta. In concert with our other metrics, GPT-3.5 has a clear advantage over GPT-Neo, aligning more closely with the reference.

Overall, GPT-3.5 appears to significantly outperform GPT-Neo across all quantitative measures. However, it is worth noting that GPT-3.5 incurs a higher cost and generally has a higher latency than GPT-Neo. In this case, the StyleSprint team would conduct a qualitative analysis to accurately determine whether the GPT-Neo descriptions do not align with brand guidelines and expectations, therefore making the cost of using the better model worthwhile. As discussed, the trade-off here is not clear-cut. StyleSprint must carefully consider that although using a commodity such as GPT-3.5 does not incur computational costs directly, on-demand costs could increase significantly as model usage rises.

The contrasting strengths of the two models pose a decision-making challenge. While one clearly excels in performance metrics and alignment with CLIP, implying higher accuracy and semantic correctness, the other is significantly more resource-efficient and scalable, which is crucial for cost-effectiveness. At this stage, it becomes critical to assess model outcomes qualitatively and to engage stakeholders to help understand organizational priorities.

With these considerations in mind, we'll revisit qualitative considerations such as transparency, bias, and fairness and how they play into the broader picture of deploying a responsible and effective AI system.

Responsible AI considerations

Addressing implicit or covert societal biases in AI systems is crucial to ensure responsible AI deployment. Although it may not seem obvious how a simple product description could introduce bias, the language used can inadvertently reinforce stereotypes or exclude certain groups. For instance, descriptions that consistently associate certain body types or skin tones with certain products or that unnecessarily default to gendered language can unintentionally perpetuate societal biases. However, with a structured mitigation approach, including algorithmic audits, increased model transparency, and stakeholder engagement, StyleSprint can make sure its brand promotes equity and inclusion.

Addressing and mitigating biases

We present several considerations, as suggested by Costanza-Chock et al. in *Who Audits the Auditors? Recommendations from a field scan of the algorithmic auditing ecosystem*:

- **Professional environment examination**: Creating a supportive professional environment is crucial for addressing algorithmic fairness. Implementing whistleblower protections facilitates the safe reporting of biases and unfair practices while establishing processes for individuals to report harms to ensure these concerns are addressed proactively.

- **Custom versus standardized audit frameworks**: While custom audit frameworks are expected, considering standardized methods may enhance rigor and transparency in bias mitigation efforts. Engaging with external auditing entities could offer unbiased evaluations of StyleSprint's AI systems, aligning with the observations by Costanza-Chock et al. (2022).

- **Focusing on equity, not just equality**: Equity notions acknowledge differing needs, essential for a comprehensive approach to fairness. Performing intersectional and small population analyses could help you to understand and address biases beyond legally protected classes.

- **Disclosure and transparency**: Disclosing audit methods and outcomes can foster a culture of transparency and continuous improvement. Officially released audits could help you establish best practices and gain stakeholder trust.

- **Mixed methods analyses**: As presented, a mix of technical and qualitative analyses could provide a holistic view of the system's fairness. Engaging non-technical stakeholders could emphasize qualitative analyses.

- **Community and stakeholder engagement**: Again, involving diverse groups and domain experts in audits could ensure diverse perspectives are considered in bias mitigation efforts. Establishing feedback loops with stakeholders could facilitate continuous improvement.

- **Continuous learning and improvement**: Staying updated on emerging standards and best practices regarding AI fairness is crucial for continuous improvement. Fostering a culture of learning could help in adapting to evolving fairness challenges and regulatory landscapes, thus ensuring StyleSprint's AI systems remain fair and responsible over time.

Transparency and explainability

Generally, explainability in machine learning refers to the ability to understand the internal mechanics of a model, elucidating how it makes decisions or predictions based on given inputs. However, achieving explainability in generative models can be much more complex. As discussed, unlike discriminative machine learning models, generative models do not have the objective of learning a decision boundary, nor do they reflect a clear notion of features or a direct mapping between input features and predictions. This absence of feature-based decision-making makes traditional explainability techniques ineffective for generative foundational models such as GPT-4.

Alternatively, we can adopt some pragmatic transparency practices, such as clear documentation made accessible to all relevant stakeholders, to foster a shared understanding and expectations regarding the model's capabilities and usage.

The topic of explainability is a critical space to watch, especially as generative models become more complex and their outcomes become increasingly more difficult to rationalize, which may present unknown risk implications.

Promising research from Anthropic, OpenAI, and others suggests that sparse autoencoders—neural networks that activate only a few neurons at a time—could facilitate the identification of abstract and understandable patterns. This method could help explain the network's behavior by highlighting features that align with human concepts.

Final deployment

Assuming we have carefully gathered quantitative and qualitative feedback regarding the best model for the job, we can select our model and update our production environment to deploy and serve it. We will continue to use FastAPI for creating a web server to serve our model, and Docker to containerize our application. However, now that we have been introduced to the simplicity of LangChain, we will continue to leverage its simplified interface. Our existing CI/CD pipeline will ensure streamlined automatic deployment and continuous application monitoring. This means that deploying our model is as simple as checking-in our latest code. We begin with updating our dependencies list:

1. **Update the requirements**: Update the requirements.txt file in your project to include the necessary libraries:

   ```
   fastapi==0.68.0
   uvicorn==0.15.0
   openai==0.27.0
   langchain==0.1.0
   ```

2. **Update the Dockerfile**: Modify your Dockerfile to ensure it installs the updated requirements and properly sets up the environment for running LangChain with FastAPI:

   ```
   # Use an official Python runtime as a base image
   FROM python:3.8-slim-buster

   # Set the working directory in the container to /app
   WORKDIR /app

   # Copy the current directory contents into the container at /app
   COPY . /app

   # Install any needed packages specified in requirements.txt
   RUN pip install --no-cache-dir -r requirements.txt

   # Make port 80 available to the world outside this container
   EXPOSE 80

   # Define environment variable
   ENV NAME World

   # Run app.py when the container launches
   CMD ["uvicorn", "app:app", "--host", "0.0.0.0", "--port", "80"]
   ```

3. **Update the FastAPI application**: Modify your FastAPI application to utilize Langchain for interacting with GPT-3.5. Ensure your OpenAI API key is securely stored and accessible to your application:

   ```
   from fastapi import FastAPI, HTTPException, Request
   from langchain.llms import OpenAI
   import os

   # Initialize FastAPI app
   ```

```
app = FastAPI()

# Setup Langchain with GPT-3.5
llm = OpenAI(model _ name='text-davinci-003',
             temperature=0.7,
             max _ tokens=256,
             api _ key=os.environ['OPENAI _ API _ KEY'])

@app.post("/generate/")
async def generate _ text(request: Request):
    data = await request.json()
    prompt = data.get('prompt')
    if not prompt:
        raise HTTPException(status _ code=400,
            detail="Prompt is required")
    response = llm(prompt)
    return {"generated _ text": response}
```

Testing and monitoring

Once the model is deployed, perform necessary tests to ensure the setup works as expected. Continue to monitor the system's performance, errors, and other critical metrics to ensure reliable operation.

By this point, we have updated our production environment to deploy and serve GPT-3.5, facilitating the generation of text based on the prompts received via the FastAPI application. This setup ensures a scalable, maintainable, and secure deployment of our new generative model. However, we should also explore some best practices regarding application reliability.

Maintenance and reliability

Maintaining reliability in our StyleSprint deployment is critical. As we employ Langchain with FastAPI, Docker, and CI/CD, it's essential to set up monitoring, alerting, automatic remediation, and failover mechanisms. This section outlines a possible approach to ensure continuous operation and robustness in our production environment:

- **Monitoring tools**: Integrate monitoring tools within the CI/CD pipeline to continuously track system performance and model metrics. This step is fundamental for identifying and rectifying issues proactively.

- **Alerting mechanisms**: Establish alerting mechanisms to notify the maintenance team whenever anomalies or issues are detected. Tuning the alerting thresholds accurately is crucial to catch issues early and minimize false alarms.

- **Automatic remediation**: Utilize Kubernetes' self-healing features and custom scripts triggered by certain alerts for automatic remediation. This setup aims to resolve common issues autonomously, reducing the need for human intervention.

- **Failover mechanisms**: Implement a failover mechanism by setting up secondary servers and databases. In case of primary server failure, these secondary setups take over to ensure continuous service availability.

- **Regular updates via CI/CD**: Employ the CI/CD pipeline for managing, testing, and deploying updates to LangChain, FastAPI, or other components of the stack. This process keeps the deployment updated and secure, reducing the maintenance burden significantly.

By meticulously addressing each of these areas, you'll be laying down a solid foundation for a reliable and maintainable StyleSprint deployment.

Summary

This chapter outlined the process of transitioning the StyleSprint generative AI prototype to a production-ready deployment for creating engaging product descriptions on an e-commerce platform. It started with setting up a robust Python environment using Docker, GitHub, and CI/CD pipelines for efficient dependency management, testing, and deployment. The focus then shifted to selecting a suitable pretrained model, emphasizing alignment with project goals, computational considerations, and responsible AI practices. This selection relied on both quantitative benchmarking and qualitative evaluation. We then outlined the deployment of the selected model using FastAPI and LangChain, ensuring a scalable and reliable production environment.

Following the strategies outlined in this chapter will equip teams with the necessary insights and steps to successfully transition their generative AI prototype into a maintainable and value-adding production system. In the next chapter, we will explore fine-tuning and its importance in LLMs. We will also weigh in on the decision-making process, addressing when it is more beneficial to fine-tune versus zero or few-shot prompting.

Part 2:
Practical Applications of
Generative AI

This part focuses on the practical applications of generative AI, including fine-tuning models for specific tasks, understanding domain adaptation, mastering prompt engineering, and addressing ethical considerations. It aims to provide hands-on insights and methodologies for effectively implementing and leveraging generative AI in various contexts with a focus on responsible adoption.

This part contains the following chapters:

- *Chapter 5, Fine-Tuning Generative Models for Specific Tasks*
- *Chapter 6, Understanding Domain Adaptation for Large Language Models*
- *Chapter 7, Mastering the Fundamentals of Prompt Engineering*
- *Chapter 8, Addressing Ethical Considerations and Charting a Path toward Trustworthy Generative AI*

Fine-Tuning Generative Models for Specific Tasks

In our narrative with StyleSprint, we described using a pre-trained generative AI model for creating engaging product descriptions. While this model showed adeptness in generating diverse content, StyleSprint's evolving needs require a shift in focus. The new challenge is not just about producing content but also about engaging in specific, task-oriented interactions such as automatically answering customer's specific questions about the products described.

In this chapter, we introduce the concept of fine-tuning, a vital step in adapting a pre-trained model to perform specific downstream tasks. For StyleSprint, this means transforming the model from a versatile content generator to a specialized tool capable of providing accurate and detailed responses to customer questions.

We will explore and define a range of scalable fine-tuning techniques, comparing them with other approaches such as in-context learning. We will demonstrate advanced fine-tuning methods, including parameter-efficient fine-tuning and prompt tuning, to demonstrate how they can fine-tune a model's abilities for specific tasks such as Q&A.

By the end of this chapter, we will have trained a language model to answer questions and do so in a way that aligns with StyleSprint's brand guidelines. However, before we explore the mechanics of fine-tuning and its importance in our application, we will revisit the history of fine-tuning in the context of LLMs.

Foundation and relevance – an introduction to fine-tuning

Fine-tuning is the process of leveraging a model pre-trained on a large dataset and continuing the training process on a smaller, task-specific dataset to improve its performance on that task. It may also involve additional training that adapts a model to the nuances of a new domain. The latter is known as domain adaptation, which we will cover in *Chapter 6*. The former is typically referred to as task-specific fine-tuning, and it can be performed to accomplish several tasks, including Q&A, summarization, classification, and many others. For this chapter, we will focus on task-specific fine-tuning to improve a general-purpose model's performance when answering questions.

For StyleSprint, fine-tuning a model to handle a specific task such as answering customer inquiries about products introduces unique challenges. Unlike generating product descriptions, which primarily involves language generation using an out-of-the-box pre-trained model, answering customer questions requires the model to have an extensive understanding of product-specific data and should have a brand-aware voice. Specifically, the model must accurately interpret and respond to questions about product features, sizes, availability, user reviews, and many other details. It should also produce answers consistent with StyleSprint's distinct brand tone. This task requires both generalized natural language proficiency (from pre-training) and robust knowledge of product metadata and customer feedback, accomplished through fine-tuning.

Models such as GPT initially learn to predict text through an unsupervised learning process that involves being trained on wide-ranging and vast datasets. This pre-training phase exposes the model to a diverse array of texts, enabling it to gain a broad understanding of language, including syntax, grammar, and context, without any specific task-oriented guidance. However, fine-tuning applies task-oriented, supervised learning to refine the model's capabilities to accomplish the specified task – specifically, semi-supervised learning, which, as described by Radford et al. (2018), involves adapting the model to a specific supervised task by exposing it to a dataset comprising input sequences ($x1, ..., xm$) and corresponding labels (y).

Throughout the chapter, we will detail the fine-tuning process, including how to selectively train the model on a curated dataset of product-related information and customer interactions, enabling it to respond with the informed, brand-aligned precision that customers expect. However, fine-tuning an LLM, which could have billions of parameters, would typically require an enormous number of resources and time. This is where advanced techniques such as **Parameter-Efficient Fine-Tuning (PEFT)** become particularly valuable in making fine-tuning accessible.

PEFT

Traditional fine-tuning methods become increasingly impractical as the model size grows due to the immense computational resources and time required to train and update all model parameters. For most businesses, including larger organizations, a classical approach to fine-tuning is cost-prohibitive and, effectively, a non-starter.

Alternatively, PEFT methods modify only a small subset of a model's parameters, reducing the computational burden while still achieving state-of-the-art performance. This method is advantageous for adapting large models to specific tasks without extensive retraining.

One such PEFT method is the **Low-Rank Adaptation (LoRA)** methodology, developed by Hu et al. (2021).

LoRA

The LoRA method focuses on selectively fine-tuning specific components within the Transformer architecture to enhance efficiency and effectiveness in LLMS. LoRA targets the weight matrices found in the self-attention module of the Transformer, which, as discussed in *Chapter 3*, are key to its functionality and include four matrices: w_q (query), w_k (key), w_v (value), and w_o (output). Although these matrices can be divided into multiple heads in a multi-head attention setting – where each *head* represents one of several parallel attention mechanisms that process inputs independently – LoRA treats them as singular matrices, simplifying the adaptation process.

LoRA's approach involves adapting only the attention weights for downstream tasks, while the weights in the other component of the Transformer, the **feed-forward network (FFN)**, are unchanged. This decision to focus exclusively on the attention weights and freeze the FFN is made for simplicity and parameter efficiency. By doing so, LoRA ensures a more manageable and resource-efficient fine-tuning process, avoiding the complexities and demands of retraining the entire network.

This selective fine-tuning strategy enables LoRA to effectively tailor the model for specific tasks while maintaining the overall structure and strengths of the pre-trained model. This makes LoRA a practical solution for adapting LLMs to new tasks with a reduced computational burden without requiring comprehensive parameter updates across the entire model (Liu et al., 2021).

Building upon the foundation of LoRA, **Adaptive Low-Rank Adaptation (AdaLoRA)**, as introduced in a study by Liu et al. (2022), represents a further advancement in PEFT methods. The key difference between LoRA and AdaLoRA lies in (as the name suggests) its adaptiveness. While LoRA applies a consistent, low-rank approach to fine-tuning across the model, AdaLoRA tailors the updates to the needs of each layer, offering a more flexible and potentially more effective way to fine-tune large models for specific tasks.

AdaLoRA

AdaLoRA's key innovation lies in its adaptive allocation of the **parameter budget** among the weight matrices of the pre-trained model. Many PEFT methods tend to distribute the parameter budget evenly across all pre-trained weight matrices, potentially neglecting the varying importance of different weight parameters. AdaLoRA overcomes this by assigning importance scores to these weight matrices and allocating the parameter budget accordingly. **Importance scores** in the context of AdaLoRA are metrics used to determine the significance (or importance) of different weight parameters in a model, guiding the allocation of the parameter budget more effectively during fine-tuning.

> **Note**
> *Parameter budget* refers to the predefined limit on the number of additional parameters that can be introduced during the fine-tuning of a pre-trained model. This budget is set to ensure that the model's complexity does not increase significantly, which can lead to challenges such as overfitting, increased computational costs, and longer training times.

Additionally, AdaLoRA applies **singular value decomposition (SVD)** to efficiently organize the incremental updates made during the model's fine-tuning process. SVD allows for the effective pruning of singular values associated with less critical updates, reducing the overall parameter budget required for fine-tuning. It is important to note that this method also avoids the need for computationally intensive exact computations, making the fine-tuning process more efficient.

AdaLoRA has been empirically tested across various domains, including natural language processing, question-answering, and natural language generation. Extensive experiments have demonstrated its effectiveness in improving model performance, particularly in question-answering tasks. The adaptability and efficiency of AdaLoRA make it an ideal choice for applications requiring precise and efficient model adjustments for complex tasks.

In the case of StyleSprint, AdaLoRA presents an opportunity to fine-tune its language model for answering customer questions without the considerable overhead that would be incurred by traditional fine-tuning, which would require adjusting all of the model parameters. By adopting AdaLoRA, StyleSprint can efficiently adapt its model to handle nuanced customer inquiries by adjusting significantly fewer parameters. Specifically, AdaLoRA's adaptive allocation of parameter budgets means that StyleSprint can optimize its model for the specific nuances of customer queries without using extensive computational resources.

By the end of this chapter, we will have fine-tuned an LLM using AdaLoRA for our Q&A task. However, we should first decide whether fine-tuning is truly the right approach. Prompt-based LLMs offer a viable alternative known as in-context learning, where the model can learn from examples given in the prompt, meaning that the prompt would contain the customer's question paired with a few key historical examples of how other questions were answered. The model can infer from the examples how to answer the question at hand in a way that is consistent with the examples. In the next section, we will explore the benefits and drawbacks of in-context learning to help us determine whether fine-tuning is the best approach to enable a model to answer very specific questions.

In-context learning

In-context learning is a technique where the model generates responses based on a few examples provided in the input prompt. This method leverages the model's pre-trained knowledge and the specific context or examples included in the prompt to perform tasks without the need for parameter updates or retraining. The general approach, detailed in *Language Models are Few-Shot Learners* by Brown et al. (2020), describes how the extensive pre-training of these models enables them to perform tasks and generate responses based on a limited set of examples paired with instructions embedded within prompts. Unlike traditional methods that require fine-tuning for each specific task, in-context learning allows the model to adapt and respond based on the additional context provided at inference.

Central to in-context learning is the concept of few-shot prompting, which is critical for enabling models to adapt to and perform tasks without additional training data, relying instead on their pre-trained knowledge and the context provided within input prompts. For context, we'll describe how an LLM typically works, which is known as the zero-shot approach, and contrast it to in-context learning, which uses the few-shot approach:

- **Zero-shot prompting**: Models such as GPT respond to instruction based on their vast pre-training and the specific task or instruction described in the input prompt. These models estimate a conditional probability distribution over possible outputs for a given input sequence, x. The model calculates the likelihood of a potential output sequence, y, expressed as $P(y|x)$. This computation is performed without prior examples specific to the task, relying entirely on the model's general pre-training. Meaning, the zero-shot approach has no specific context apart from its general knowledge. For example, if we were to ask *Are winter coats available in children's sizes?*, the model could not provide a specific answer about StyleSprint's inventory. It could only provide some generic answer.

- **Few-shot prompting**: Using the few-shot approach, we provide the model with a prompt paired with a few examples. These examples are concatenated to the prompt (represented as x) to form an extended input sequence. So, our question *Are winter coats available in children's sizes?* might be paired with a few examples such as the following:

  - **Q**: Do you sell anything in children's sizes?

    A: Any items for children are specifically listed on the "StyleSprint for Kids" page.

  - **Q**: What do you offer for kids?

    A: StyleSprint offers a variety of children's fashions on its "StyleSprint for Kids" page.

The LLM then computes the probability of generating a specific output sequence, y, given this extended input sequence, x. Mathematically, this can be conceptualized as the model estimating the joint probability distribution of y and x (where x includes both the prompt and the few-shot examples, as demonstrated previously). The model uses this joint probability distribution to generate a response consistent with the instructions paired with the examples given in the input sequence.

In both cases, the model's ability to adapt its output based on the given context, whether with zero examples or a few, demonstrates the flexibility and sophistication of its underlying architecture and training. However, the few-shot approach allows the LLM to learn from the very specific examples provided.

Let's consider how StyleSprint could apply in-context learning to answer customer queries. Performance using in-context learning (or the few-shot approach) consistently reflects significant gains over zero-shot behavior (Brown et al., 2020). We can expand our prior example to where a customer asks about the availability of a specific product. Again, the StyleSprint team could systematically append a few examples to each prompt as follows.

Here is the prompt: Respond to the following {question} about product availability.

These are some examples:

- Example 1:

  - **Customer query**: `Do you carry black leather handbags?`

  - **AI response**: `Give me a moment while I retrieve information about that particular item.`

- Example 2:

  - **Customer query**: `Do you have the silk scarves in blue?`

  - **AI response**: `Let me search our inventory for blue silk scarves.`

StyleSprint can provide examples that effectively help the model understand the nature of the inquiry and generate a response that is informative and aligned with the company's policies and product offerings. In this example, we see that the responses are intended to be paired with a search component. This is a common approach and can be accomplished using a technique called **Retrieval Augmented Generation** (**RAG**), which is a component that facilitates retrieval of real-time data to inform the generated response. Combining a few-shot in-context learning approach with RAG could ensure that the system provides a logical and specific answer.

In-context learning using a few-shot approach allows the model to rapidly adapt to various customer queries using a limited set of examples. When augmented with RAG, StyleSprint could potentially satisfy their use case and reduce the time and resources needed to fine-tune. However, this approach must be weighed against the depth of specialization and consistency of task-specific fine-tuning, which, as described, could also produce highly accurate answers that fit the brand tone.

In the next section, we will formulate metrics that help us draw a direct comparison to guide StyleSprint in making an informed decision that best suits its customer service objectives and operational framework.

Fine-tuning versus in-context learning

We learned how in-context learning could allow StyleSprint's model to handle a diverse range of customer queries without requiring extensive retraining. Specifically, a few-shot approach combined with RAG could facilitate quick adaptation to new inquiries, as the model can generate responses based on a few examples. However, the effectiveness of in-context learning heavily relies on the quality and relevance of the examples provided in the prompts. Its success would also rely on the implementation of RAG. Moreover, without fine-tuning, responses may lack consistency or may not adhere as strictly to StyleSprint's brand tone and customer service policies. Finally, depending entirely on a generative model without fine-tuning may inadvertently introduce bias, as discussed in *Chapter 4*.

In practice, we have two very comparable and viable approaches. However, to make an informed decision, we should first perform a more in-depth comparison using quantitative methods.

To impartially assess the efficacy of in-context learning compared to fine-tuning, we can measure the quality and consistency of the generated responses. We can accomplish this using established and reliable metrics to compare outcomes from each of the approaches. Like prior evaluations, we will want to apply quantitative and qualitative methods applied across the following key dimensions:

- **Alignment with human judgment**: We can again apply semantic similarity to provide a quantitative measure of how often the model's responses are correct or relevant based on a reference answer written by a human.

  StyleSprint's brand communication experts can review a subset of the responses to provide a qualitative evaluation of the response accuracy and alignment with brand tone and voice.

- **Consistency and stability**: It is important to measure the degree to which questions are answered consistently each time despite minor variations in how the question is posed. Again, we can leverage semantic similarity to compare each new output to the prior when the input is held constant.

In addition to evaluating the quality of model responses for each approach, we can also directly compare the operational and computational overhead required for each.

For fine-tuning, we will need to understand the overhead involved in training the model. While the PEFT method will significantly reduce the training effort, there could be considerably more infrastructure-related costs compared to in-context learning, which requires no additional training. Alternatively, for in-context learning, commoditized models such as OpenAI's GPT-4 have a per-token cost model. StyleSprint must also consider the cost of tokens required to embed a sufficient number of few-shot examples in the prompt.

In both cases, StyleSprint will incur some operational costs to create best-in-class examples written by humans that can be used as a "gold standard" in either the few-shot approach or for additional model training.

By conducting these comparative tests and analyzing the results, StyleSprint will gain valuable insights into which approach – in-context learning or fine-tuning – best aligns with its operational goals and customer service standards. This data-driven evaluation will inform the decision on the optimal AI strategy for enhancing their customer service experience. We will implement these comparisons in the practice project that follows.

Practice project: Fine-tuning for Q&A using PEFT

For our practice project, we will experiment with AdaLoRA to efficiently fine-tune a model for a customer query and compare it directly to the output of a **state-of-the-art** (**SOTA**) model using in-context learning. Like the previous chapter, we can rely on a prototyping environment such as Google Colab to complete the evaluation and comparison of the two approaches. We will demonstrate how to configure model training to use AdaLoRA as our PEFT method.

Background regarding question-answering fine-tuning

Our project utilizes the Hugging Face training pipeline library, a widely recognized resource in the machine learning community. This library offers a variety of pre-built pipelines, including one for question-answering, which allows us to fine-tune pre-trained models with minimal setup. Hugging Face pipelines abstract much of the complexity involved in model training, making it accessible for developers to implement advanced natural language processing tasks directly and efficiently In particular, this pipeline behaves as an interface to a transformer model with a specific head for question-answering tasks. Recall that when we fine-tune a transformer model, we keep the architecture of the model – including the self-attention mechanism and the transformer layers – but we train the model's parameters on a specific task, which, in this case, results in a model refined specifically to answer questions. Recall our practice project in *Chapter 3* where the resulting model was a translator; we used a translator head to accomplish translation from English to French. For this project, the "head" is aligned to learn patterns in question-answering data.

However, when using a question-answer training pipeline, it is important to understand that the model does not simply memorize question-answer pairs, it learns the connection between questions and answers. Moreover, to answer appropriately, the model cannot rely entirely on training. It also requires additional context as input to compose a relevant answer. To understand this further, we decompose the model inferencing step as follows:

1. When feeding a question to a model, we must also include context relevant to the topic.

2. The model then determines the most relevant part of the context that answers the question. It does this by assigning probability scores to each token (word or sub-word) in the context.

3. The model "thinks" of the context as a potential source for the answer and assigns each token two scores: one score for being the **start** of the answer, and another for being the **end** of the answer.

4. The token with the highest "start" score and "end" score is then chosen to form the answer **span**. The span is what is presented to the user.

To provide a concrete example, if we ask the model, Does StyleSprint have any leather jackets? and provide a context of StyleSprint sells a variety of coats, jackets and outerwear, the model will process this context and identify that the most likely answer is something like Yes, StyleSprint sells a variety of outerwear. However, if the answer to a question is not included in the provided context, the model cannot generate a reliable answer. Additionally, if the context is too unspecific, the model may provide a more generic answer. Like in-context learning, the fine-tuned approach for question-answering requires relevant context. This means that, in practice, the model must be integrated with a search component that can retrieve additional context to pair with each question.

Consider our leather jacket example. When a question is received, the system could perform a search of its knowledge base and retrieve any contextual information relevant to a leather jacket (e.g., a paragraph about outerwear). Again, since the model was trained to answer questions in a way that aligns with the brand tone, it will extract the relevant information from the context provided to formulate an appropriate answer. Not only will integration with search provide the model with the context it needs but it will also allow the model to have up-to-date and real-time information.

Additionally, we might incorporate a confidence threshold, where the model only gives an answer if it assigns a high enough probability to the start and end tokens. If the highest probability is below this threshold, we might say the model does not know, or request more information. Overall, the model efficacy relies heavily on the quality and size of the training data as well as the relevance of the context with regard to the questions posed.

Now that we have a better understanding of how fine-tuning for question-answering works and what to expect when using the question-answering pipeline from Hugging Face, we can begin to write our implementation.

Implementation in Python

First and foremost, we install the required libraries:

```
!pip install transformers peft sentence-transformers
```

Then, we import the question-answering modules from the transformers library. For our project, we will use Google's **Flan T5 (small)**, which is considered a SOTA alternative to GPT 3.5. As one of our goals continues to be to measure the performance versus efficiency trade-off, we begin with the smallest version of Flan T5, which has 80M parameters. This will enable faster training and more rapid iteration. However, please note that even a small model trained over a small number of epochs will require a high-RAM runtime environment:

```
from transformers import (
    AutoModelForQuestionAnswering, AutoTokenizer)
model_name = " google/flan-t5-small"
tokenizer = AutoTokenizer.from_pretrained(model_name)
model = AutoModelForQuestionAnswering.from_pretrained(model_name)
```

With the pre-trained model instantiated, we can now configure the model to adapt its training process to use AdaLoRA, which, as we've learned, is specifically designed to allocate the parameter budget efficiently during the fine-tuning process:

```
from peft import AdaLoraConfig
# Example configuration; adjust parameters as needed
adapter_config = AdaLoraConfig(target_r=16)
model.add_adapter(adapter_config)
```

As discussed, fine-tuning relies heavily on the quality and size of the training data. In the StyleSprint scenario, the company could aggregate question-answer pairs from its FAQ page, social media, and customer service transcripts. For this exercise, we will construct a simple dataset that looks similar to the following:

```
demo_data = [{
"question": "What are the latest streetwear trends available at
Stylesprint?",
  "answer": "Stylesprint's latest streetwear collection includes
hoodies, and graphic tees, all inspired by the latest hip-hop fashion
trends."
...
}]
```

However, in order to integrate our dataset with the question-answer pipeline, we should first understand the `Trainer` class. The `Trainer` class in the Hugging Face transformers library expects the training and evaluation datasets to be in a specific format, usually as a PyTorch `Dataset` object, not just as simple lists of dictionaries. Further, each entry in the dataset needs to be tokenized and structured with the necessary fields such as `input_ids`, `attention_mask`, and, for question-answering tasks, `start_positions` and `end_positions`. Let us explore these in more detail:

- `input_ids`: This is a sequence of integers that represent the input sentence in the model. Each word or sub-word in the sentence is converted into a unique integer or ID. Recall from earlier chapters that this process is known as **tokenization**. The words or tokens are looked up in the vocabulary of the language model and the corresponding integer is then used in the model. For example, a sentence such as *I love Paris* might be converted into something like `[101, 354, 2459]`.

- `attention_mask`: An attention mask is a sequence of binary values where 1s indicate real tokens and 0s indicate padding tokens. In other words, in the places where 1s are present, the model will understand that those places need attention and the places with 0s will be ignored by the model. This is crucial when dealing with sentences of varying lengths and dealing with batches of sentences in training models.

- `start_positions` and `end_positions`: These are for question-answering tasks. They represent the indices of the start and end tokens of the answer in the tokenized form of the context. For example, in the context *Paris is the capital of France*, if the question is *What is the capital of France?* and the answer given is *Paris*, after tokenization, `start_position` and `end_position` will correspond to the index of *Paris* in the context.

With that understanding, we can create a class that adapts our dataset to meet the expectations of the trainer, as follows:

```
from torch.utils.data import Dataset

class StylesprintDataset(Dataset):
    def __init__(self, tokenizer, data):
```

```
        tokenizer.pad_token = tokenizer.eos_token
        self.tokenizer = tokenizer
        self.data = data
```

For the complete custom dataset class code, visit this book's GitHub repository at `https://github.com/PacktPublishing/Generative-AI-Foundations-in-Python`.

With the training set prepared and our pipeline configured to apply the AdaLoRA method, we can finally move to the training step. For this project, we will configure the training to run for just a few epochs, but in the StyleSprint scenario, a much more robust training process would be required:

```
from transformers import Trainer, TrainingArguments

# Split the mock dataset into training and evaluation sets (50/50)
train_data = StylesprintDataset(
    tokenizer, demo_data[:len(demo_data)//2])
eval_data = StylesprintDataset(
    tokenizer, demo_data[len(demo_data)//2:])

# Training arguments
training_args = TrainingArguments(
    output_dir="./results",
    num_train_epochs=10,
    per_device_train_batch_size=16,
    per_device_eval_batch_size=64,
    warmup_steps=500,
    weight_decay=0.01,
    logging_dir="./logs",
    logging_steps=10,
)

# Initialize the Trainer
trainer = Trainer(
    model=model,
    args=training_args,
    train_dataset=train_data,
    eval_dataset=eval_data
)

# Start training
trainer.train()
```

For our simple experiment, we do not expect a highly performant model; however, we can learn how to interpret the training output, which describes how well the model performed on the evaluation samples. The `Trainer` class will output a training summary that includes the loss metric.

Training loss

Training loss is a measure of how well the model is performing; a lower loss indicates better performance. In many deep learning models, especially those dealing with complex tasks such as language understanding, it's common to start with a relatively high loss. The expectation is that this value should decrease as training progresses.

In the early stages of training, a high loss isn't a cause for alarm as it commonly decreases as the model continues to learn. However, if the loss remains high, this signals that additional training may be needed. If the loss continues to be high after prolonged training, the learning rate and other hyperparameters may require adjustment, as an inappropriate learning rate can impact the model's learning effectiveness. Moreover, the quality and quantity of your training data should be evaluated as insufficient data can hinder the training. For example, as we only use a few examples for the experiment, we expect a relatively high loss.

The next step is to use our newly fine-tuned model to infer or predict. We should also secure our trained model parameters so we can reuse it without retraining:

```python
import torch

# save parameters
model.save_pretrained("./stylesprint_qa_model")

def ask_question(model, question, context):
    # Tokenize the question and context
    inputs = tokenizer.encode_plus(question, context,
        add_special_tokens=True, return_tensors="pt")

    # Get model predictions
    with torch.no_grad():
        outputs = model(**inputs)

    # Get the start and end positions
    answer_start_scores = outputs.start_logits
    answer_end_scores = outputs.end_logits

    # Find the tokens with the highest `start` and `end` scores
    answer_start = torch.argmax(answer_start_scores)
    answer_end = torch.argmax(answer_end_scores) + 1

    # Convert the tokens to the answer string
    answer = tokenizer.convert_tokens_to_string(
        tokenizer.convert_ids_to_tokens(
            inputs["input_ids"][0][answer_start:answer_end]
```

```
            )
        )
    return answer

question = "What is the return policy for online purchases?"
context = """Excerpt from return policy returned from search."""

answer = ask_question(model, question, context)
print(answer)
```

As discussed, we introduce context along with a question to the model, so that it can identify which fragment of the context responds most appropriately to the query. Consequently, we may want to consider integrating a vector search system (such as RAG) to automatically identify relevant documents from large datasets based on semantic similarities to a query. These search results may not provide specific answers, but the trained QA model can extract more precise answers from the results.

With this hybrid approach, the vector search system first retrieves documents or text segments that are semantically related to the query. The QA model then analyzes this context to identify the precise answer that aligns with StyleSprint's guidelines and expectations.

Evaluation of results

To evaluate our model outcomes, StyleSprint might apply the qualitative and quantitative approaches we have discussed in the chapter already. For the purpose of our experiment, we can measure the output of the model to a golden standard response using a simple measure for semantic similarity:

```
from sentence_transformers import SentenceTransformer, util
import pandas as pd

# Example of a gold standard answer written by a human
gs = "Our policy at Stylesprint is to accept returns on online
purchases within 30 days, with the condition that the items are unused
and remain in their original condition."
# Example of answer using GPT 3.5 with in-context learning reusing a
relevant subset of the training data examples
gpt_35 = "Stylesprint accepts returns within 30 days of purchase,
provided the items are unworn and in their original condition."

# Load your dataset
dataset = pd.DataFrame([
    (gs, gpt_35, answer)
])# pd.read_csv("dataset.csv")
dataset.columns = ['gold_standard_response',
    'in_context_response', 'fine_tuned_response']
# Load a pre-trained sentence transformer model
```

```
eval_model = SentenceTransformer('all-MiniLM-L6-v2')

# Function to calculate semantic similarity
def calculate_semantic_similarity(model, response, gold_standard):
    response_embedding = model.encode(
        response, convert_to_tensor=True)
    gold_standard_embedding = model.encode(gold_standard,
        convert_to_tensor=True)
    return util.pytorch_cos_sim(response_embedding,
        gold_standard_embedding).item()

# Measure semantic similarity
dataset['in_context_similarity'] = dataset.apply(
    lambda row:calculate_semantic_similarity(
        eval_model, row['in_context_response'],
        row['gold_standard_response']
    ), axis=1)
dataset['fine_tuned_similarity'] = dataset.apply(
    lambda row:calculate_semantic_similarity(
        eval_model, row['fine_tuned_response'],
        row['gold_standard_response']
    ), axis=1)

# Print semantic similarity
print("Semantic similarity for in-context learning:",
    dataset['in_context_similarity'])
print("Semantic similarity for fine-tuned model:",
    dataset['fine_tuned_similarity'])
```

The results of our evaluation are as follows:

	PEFT Flan T5	GPT 3.5T
	Fine-tuned	In-context
Semantic Similarity	0.543	0.91

Table 5.1: Semantic similarity scores for fine-tuned Flan and GPT 3.5 Turbo, respectively

Undoubtedly, the in-context learning arrived at an answer that was much closer to our gold standard reference. However, the fine-tuned model was not far behind. This tells us that with a more robust training dataset and considerably more epochs, the fine-tuned model could be comparable to GPT 3.5. With more iteration and experimentation, StyleSprint could have a very robust fine-tuned model to answer very specific questions for its customers.

Summary

In this chapter, we focused on the strategic decision-making process between fine-tuning and in-context learning for StyleSprint's AI-driven customer service system. While in-context learning, particularly few-shot learning, offers adaptability and resource efficiency, it may not consistently align with StyleSprint's brand tone and customer service guidelines. This method relies heavily on the quality and relevance of the examples provided in the prompts, requiring careful crafting to ensure optimal outcomes.

On the other hand, PEFT methods such as AdaLoRA, offer a more focused approach to adapt a pre-trained model to the specific demands of customer service queries. PEFT methods modify only a small subset of a model's parameters, reducing the computational burden while still achieving high performance. This efficiency is crucial for real-world applications where computational resources and response accuracy are both key considerations.

Ultimately, the choice between in-context learning and fine-tuning is not just a technical decision but also a strategic one, deeply intertwined with the company's operational goals, resource allocation, and the desired customer experience. The chapter suggests conducting comparative tests to assess the efficacy of both approaches, evaluating outcomes at scale through reliable metrics. This data-driven evaluation will inform StyleSprint's decision on the optimal AI strategy for enhancing their customer service experience.

In summary, we now have a more complete understanding of the implications of fine-tuning versus in-context learning in LLMs, specifically in the context of customer service. It highlights the need for a company like StyleSprint to make a well-informed strategic decision, balancing the depth of specialization and consistency offered by fine-tuning against the adaptability and efficiency of in-context learning.

In the next chapter, we will explore PEFT for domain adaptation where the outcome of our training is a general-purpose model refined to understand a highly specific domain like finance or law.

References

This reference section serves as a repository of sources referenced within this book; you can explore these resources to further enhance your understanding and knowledge of the subject matter:

- Radford, A., Narasimhan, K., Salimans, T., and Sutskever, I. (2018). *Improving language understanding by generative pre-training*. OpenAI.

- Hu, E. J., Shen, Y., Wallis, P., Li, Y., Wang, S., Wang, L., and Chen, W. (2021). *LoRA: Low-Rank Adaptation of Large Language Models*. ArXiv. /abs/2106.09685

- Zhang, Q., Chen, M., Bukharin, A., He, P., Cheng, Y., Chen, W., and Zhao, T. (2023). *Adaptive Budget Allocation for Parameter-Efficient Fine-Tuning*. ArXiv. /abs/2303.10512

- Brown TB, Mann B, Ryder N, et al. 2020. *Language Models are Few-Shot Learners*. ArXiv:2005.14165.

Understanding Domain Adaptation for Large Language Models

In the previous chapter, we examined how **Parameter-Efficient Fine-Tuning (PEFT)** enhances **large language models (LLMs)** for specific tasks such as question-answering. In this chapter, we will be introduced to domain adaptation, a distinct fine-tuning approach. Unlike task-specific tuning, domain adaptation equips models to interpret language that's unique to specific industries or domains, addressing the gap in LLMs' understanding of specialized language.

To illustrate this, we'll introduce *Proxima Investment Group*, a hypothetical digital-only investment firm aiming to adapt an LLM to its specific financial language using internal data. We'll demonstrate how modifying the LLM to process the specific terminology and nuances typical in Proxima's environment enhances the model's relevance and effectiveness in the financial domain.

We'll also explore the practical steps Proxima might take, such as selecting relevant internal datasets for training, applying PEFT methods such as **Low-Rank Adaptation (LoRA)** to adapt the model efficiently, and using masking techniques to refine the model's comprehension. Then, we'll explore how Proxima can evaluate the success of this domain adaptation, assessing the model's performance in tasks such as analyzing financial trends, responding to client inquiries, and generating reports that align with Proxima's internal standards and market position.

By the end of this chapter, we will clearly understand the theoretical underpinnings of domain adaptation and its real-world application, particularly in a complex sector such as finance, where the model's depth of domain understanding can significantly impact business outcomes.

Let's begin by demystifying the concept, exploring its technical underpinnings, and discussing its importance in accomplishing domain-specific business objectives.

Demystifying domain adaptation – understanding its history and importance

In the context of generative LLMs, domain adaptation specifically tailors models such as **BLOOM**, which have been pre-trained on extensive, generalized datasets (such as news articles and Wikipedia entries) for enhanced understanding of texts from targeted sectors, including biomedical, legal, and financial fields. This type of refinement can be pivotal as LLMs, despite their vast pre-training, may not inherently capture the intricate details and specialized terminology inherent to these domains. This adaptation involves a deliberate process of realigning the model's learned patterns to the linguistic characteristics, terminologies, and contextual nuances prevalent in the target domain.

Domain adaptation operates within the ambit of **transfer learning**. In this broader paradigm, a model's learnings from one task are repurposed to improve its efficacy on a related yet distinct task. This approach capitalizes on the model's pre-learned features to improve its efficiency and accuracy on the subsequent task, markedly reducing its reliance on large volumes of domain-specific data and computational resources. Specifically, we begin with a model that's been trained on broad datasets and use it as a starting point to adapt to specialized domains thereby augmenting their accuracy, relevance, and applicability to more targeted use cases.

In practice, several methodologies can be employed to tailor the model to specific domains, including the following:

- **Continued pre-training**: The model undergoes additional pre-training on domain-specific corpora, allowing its parameters to be adapted incrementally to the target domain's linguistic features, as highlighted in research by Gururangan et al. 2020.

- **Intermediate task training**: Here, the model is trained on intermediate tasks, utilizing domain-specific data before being fine-tuned for downstream applications. This step facilitates a more robust adaptation to the domain (Pruksachatkun et al., 2020).

- **Data augmentation**: Techniques such as **back translation** (Xie et al., 2019) and **token replacement** (Anaby-Tavor et al., 2020) are leveraged to generate synthetic domain-specific training examples from limited actual data:

  - **Back translation** entails translating an existing text from one language (for example, English) into another (for example, French) and then translating it back to the original language. This process generates paraphrased versions of the original text while preserving its semantics.

  - **Token replacement** involves altering individual words within a sentence to generate new sentences. This alteration usually aims to preserve the semantic meaning of the original sentence while introducing variations.

- **Multi-task learning**: This framework concurrently optimizes the model for both generic and domain-specific tasks during the adaptation phase, as demonstrated by Clark et al. 2019.

As domain adaptation techniques evolve, they increasingly enhance model performance in specialized fields, even with reduced amounts of domain-specific data. As discussed in *Chapter 4*, more recent developments have focused on the computational efficiency of these techniques. Adaptation methods such as LoRA facilitate significant model adjustments with minimal parameter changes without requiring comprehensive retraining. It is important to note that a model's performance will always vary based on various factors like the quality of the dataset, available computational resources, and other implementation details.

Now that we have some insight into domain adaptation techniques and their focus on computational efficiency, we can apply these concepts practically. Our practice project will leverage BLOOM, a state-of-the-art, open source LLM, to demonstrate domain adaptation for the finance sector. Leveraging PEFT, we aim to fine-tune BLOOM with minimal computational resources, illustrating the practical application of these advanced adaptation methods in enhancing model performance within the finance domain.

Practice project: Transfer learning for the finance domain

This project aims to fine-tune BLOOM on a curated corpus of specific documents to imbue it with the ability to interpret and articulate concepts specific to Proxima and its products.

Our methodology is inspired by strategies for domain adaptation across various fields, including biomedicine, finance, and law. A noteworthy study conducted by Cheng et al. in 2023 called *Adapting Large Language Models via Reading Comprehension* presents a novel approach for enhancing LLMs' proficiency in domain-specific tasks. This approach repurposed extensive pre-training corpora into formats conducive to reading comprehension tasks, significantly improving the models' functionality in specialized domains. In our case, we will apply a similar but simplified approach to continued pre-training by fine-tuning the pre-trained BLOOM model using a bespoke dataset specific to Proxima, effectively continuing the model's training. This process adjusts the model parameters incrementally to ensure that it understands the language unique to Proxima's products and offerings better.

Training methodologies for financial domain adaptation

Four our continued training strategy, we'll employ **causal language modeling** (**CLM**). This approach is part of a broader set of training methodologies that optimize model performance for various objectives. Before moving to implementation, let's try to disambiguate our chosen approach from other popular strategies to better understand the CLM methodology:

- **Masked Language Modeling**(**MLM**): A cornerstone of Transformer-based models such as BERT, MLM randomly masks parts of the input text and challenges the model to predict the masked tokens. By considering the entire context around the mask (both before and after), MLM enables a model to develop a bidirectional understanding of language, enriching its grasp of context and semantics.

- **Next-Sentence Prediction**(NSP): This methodology further broadens a model's narrative understanding by training it to discern whether two sentences logically follow each other. NSP is instrumental in teaching models about text structure and coherence, enabling them to construct and comprehend logical sequences within larger bodies of text.

- **CLM**: Our chosen path for BLOOM's adaptation diverges here, embracing CLM for its focused, sequential prediction capabilities. Unlike MLM, which looks both ways (before and after the masked token), CLM adopts a unidirectional approach, predicting each subsequent token based solely on the preceding context. This method is intrinsically aligned with natural language generation, making it especially suitable for crafting coherent, contextually rich narratives in the target domain.

In selecting CLM for BLOOM's adaptation, we'll extend the model's generative capabilities to produce text sequences that are not only logically structured but also deeply embedded with the nuance of the target domain. CLM's unidirectional nature ensures that each token that's generated is informed by a cohesive understanding of the preceding text, enabling the model to generate detailed, accurate, and domain-specific texts.

Once fine-tuning is complete, we can evaluate the efficacy of the domain-adapted BLOOM model based on its proficiency in generating contextually relevant and domain-specific narratives. We'll compare the adapted model's performance against the original model with a special focus on the model's fluency, accuracy, and overall comprehension of the target domain.

As we've done previously, we'll leverage Google Colab for our initial prototyping phase. As *Chapters 4 and 5* described, Google Colab offers a preconfigured environment that simplifies the process of testing our methodologies before we consider promoting them to production environments. All the code in this chapter is available in the Chapter 6 folder of this book's GitHub repository (https://github.com/PacktPublishing/Generative-AI-Foundations-in-Python).

We'll begin with the initial setup, which involves loading a smaller variation of **BLOOM-1b1** using the Transformers library. We'll also import the methods that we'll need to apply PEFT. For this example, we'll rely on a few libraries that can be installed as follows:

```
pip install sentence-transformers transformers peft datasets
```

Once installed, we can begin importing:

```
from transformers import (
    AutoTokenizer, AutoModelForCausalLM)
from peft import AdaLoraConfig, get_peft_model
```

The next step is to load the tokenizer and model:

```
tokenizer = AutoTokenizer.from_pretrained("bigscience/bloom-1b1")
model = AutoModelForCausalLM.from_pretrained(
    "bigscience/bloom-1b1")
```

As discussed previously, we're incorporating PEFT for efficient adaptation:

```
adapter_config = AdaLoraConfig(target_r=16)
model.add_adapter(adapter_config)
```

The PEFT technique, specifically through `AdaLoraConfig`, allows us to introduce a compact, efficient layer so that we can adapt the model to new contexts – here, the finance domain – with a significantly reduced number of trainable parameters:

```
model = get_peft_model(model, adapter_config)
model.print_trainable_parameters()
```

We must integrate the adapter to finalize the PEFT model setup, effectively creating a model variant that's optimized for our domain-specific training while focusing on efficiency. We can quantify this by examining the number of trainable parameters our model will use:

```
trainable params: 1,769,760 || all params: 1,067,084,088 ||
trainable%: 0.1658500974667331
```

The preceding code provides us with the following information:

- **Trainable parameters**: 1,769,760

- **Total parameters in the model**: 1,067,084,088

- **Percentage of trainable parameters**: 0.166%

This means that out of over 1 billion parameters in the BLOOM-1b1 model, only about 1.77 million parameters are being fine-tuned for the finance domain adaptation. This small percentage (0.166%) of trainable parameters highlights the efficiency of PEFT, allowing significant model adaptability with minimal adjustments. This is crucial for practical applications as it reduces both computational costs and the time required for training.

Next, we'll move on to preparing the data. We'll assume we have assembled texts encompassing the breadth of knowledge about specialized Proxima products and offerings such as the **Proxima Passkey**. CLM training requires distinct testing and training phases to evaluate the model's ability to accurately predict the next token in a sequence. This ensures it generalizes well beyond the training data to unseen text. During training, the loss calculation measures the difference between the model's predicted token probabilities and the actual tokens. It guides the model to adjust its parameters to minimize this loss, improving its predictive accuracy over iterations. As such, we must define training and testing texts as our dataset. An example dataset is included in this book's GitHub repository (linked earlier in the chapter).

```
dataset = load_dataset("text",
    data_files={"train": "./train.txt",
        "test": "./test.txt"}
    )
```

Next, we must apply preprocessing and tokenization. Texts are cleaned, standardized, and then converted into a numerical format (**tokens**) that the model can process. We must also truncate or pad texts to fit the model's input size constraints and prepare labels for CLM training, where the model learns to predict each subsequent token. Truncation and padding are preprocessing steps that are used to standardize the length of input texts for machine learning models, particularly those with fixed input size constraints like many language models. **Truncation** removes parts of the text to shorten inputs that exceed the model's maximum length, ensuring they fit within the specified size limit. **Padding** adds filler values (often zeros) to shorter inputs to extend them to the required length, allowing for consistent input dimensions across the dataset. Consistent input dimensions are necessary to ensure uniformity in matrix operations and computations across the entire dataset since LLMs, like other models that rely on deep learning, process inputs through layers of functions that require fixed-size vectors or matrices. In this case, we'll set the sequence length to a maximum of 512 tokens so that it aligns with the model's architecture:

```
def preprocess_function(examples):
    inputs = tokenizer(examples["text"], truncation=True,
        padding="max_length", max_length=512)
    inputs["labels"] = inputs["input_ids"].copy()
    return inputs
```

The `TrainingArguments` class configures the training process, setting parameters such as the batch size, number of epochs, and the directory for saving model checkpoints. This configuration is crucial for efficient learning and model evaluation. Meanwhile, the `Trainer` class orchestrates the model's training process. Again, continued training gradually adapts the model's parameters to generate and understand text related to the Proxima Passkey:

```
from transformers import Trainer, TrainingArguments
training_args = TrainingArguments(
    output_dir="./model_output",
    per_device_train_batch_size=2,
    num_train_epochs=5,
    logging_dir='./logs',
    logging_steps=10,
    load_best_model_at_end=True,
    prediction_loss_only=True,
)

trainer = Trainer(
    model=model,
    args=training_args,
    train_dataset=tokenized_datasets["train"],
    eval_dataset=tokenized_datasets["test"],
)
```

```
trainer.train()
model.save_pretrained("./proxima_da_model")
```

Generally, our configuration specifies the training parameters and initializes the `Trainer` class while focusing on domain adaptation. The `TrainingArguments` class is tailored to manage the training process efficiently, including logging and model-saving strategies. Remember that the batch size we choose for training the model balances the GPU's memory capacity and how quickly the model learns from the dataset. A larger batch size allows more data to be processed at once, speeding up training but requiring more memory, which can be a limitation if the GPU has restricted capacity. Conversely, a smaller batch size means the model updates its weights more frequently with fewer samples, which can benefit learning but results in slower overall progress through the dataset.

With training complete, we can use the adapted model to generate text based on prompts related to the Proxima Passkey. The model considers the prompt, generates a sequence of tokens representing the continuation, and then decodes this sequence back into human-readable text:

```
def predict(model, prompt="The Proxima Passkey is"):
    inputs = tokenizer(prompt, return_tensors="pt")
    output = model.generate(**inputs, max_length=50)
    return tokenizer.decode(output[0], skip_special_tokens=True)
```

Notice the `model.generate()` function, which takes tokenized input and produces a sequence of tokens as output. These tokens are then decoded into text.

In this example, we adapted the BLOOM language model so that it specializes in the finance domain. This involved loading the pre-trained model, applying a PEFT adapter for efficient domain adaptation, and preparing a financial dataset for model training through standardization and tokenization. After fine-tuning BLOOM with this domain-specific data, we used the model to generate text relevant to the finance sector. The final step is to evaluate this adapted model's performance compared to the original pre-trained version, focusing on its effectiveness in accurately handling financial language and concepts.

Evaluation and outcome analysis – the ROUGE metric

Quantitative and qualitative evaluations are essential to assess the adapted BLOOM model against the original, especially in the context of Proxima's language. Quantitatively, the model's output is compared against a reference dataset that mirrors Proxima's product language using the **ROUGE** metric. This comparison helps measure the overlap in key terms and styles. Additionally, it's beneficial to develop specific metrics for evaluating the model's proficiency in terms of financial terminology and concepts relevant to Proxima:

```
from rouge import Rouge
# Example reference text (what we expect the model to generate after
training on a complete dataset)
```

```
reference = "Proxima's Passkey enables seamless integration of
diverse financial portfolios, offering unparalleled access to global
investment opportunities and streamlined asset management."

# Example predicted model output
predicted = "The Proxima Passkey provides a unified platform for
managing various investment portfolios, granting access to worldwide
investment options and efficient asset control."

# Initialize the Rouge metric
rouge = Rouge()

# Compute the Rouge scores
scores = rouge.get_scores(predicted, reference)

print(scores)
```

The ROUGE score would be calculated by comparing the two texts in this example. The score measures the overlap between the predicted output and the reference text in terms of **n-grams** (sequences of words). For instance, **ROUGE-N** (where *N* can be 1, 2, or L) calculates the overlap of n-grams between the predicted and reference texts:

- **ROUGE-1** evaluates the overlap of unigrams (individual words) between the predicted and reference texts

- **ROUGE-2** assesses the overlap of bigrams (two-word phrases) between the texts

- **ROUGE-L** focuses on the longest common subsequence, which is useful for evaluating sentence-level structure similarity

The ROUGE scores range from 0 to 1 and quantify the similarity between the predicted text and a reference text, providing insights into how well a model's output matches the expected content. Scores closer to 1 indicate higher similarity or overlap, while scores near 0 suggest little to no commonality. These scores are divided into three key components – precision, recall, and the F1 score:

- **Precision** measures the proportion of words in the predicted text that are also found in the reference text. A high precision score indicates that most of the words generated by the model are relevant and appear in the reference, signifying accuracy in the model's output.

- **Recall** assesses the proportion of words from the reference text that are captured in the model's prediction. High recall implies that the model effectively includes most of the relevant content from the reference in its output, indicating comprehensiveness.

- The **F1 score** is the harmonic mean of precision and recall, balancing the two. It is especially useful for understanding the model's overall accuracy in generating text that is both relevant (precision) and comprehensive (recall). The F1 score is crucial when equal importance is given to precision and recall in evaluating the model's performance.

- Here's the output:

Metric	Recall (r)	Precision (p)	F1 Score (f)
ROUGE-1	0.35	0.333	0.341
ROUGE-2	0.053	0.048	0.05
ROUGE-L	0.35	0.333	0.341

Table 6.1: ROUGE metric outcomes

These scores indicate a moderate level of unigram overlap (ROUGE-1) between the texts but a significantly lower bigram overlap (ROUGE-2). The similarity between the ROUGE-1 and ROUGE-L scores suggests the model captures individual key terms to some extent but may struggle with longer phrase structures, pointing to areas for model improvement.

Overall, while the model demonstrates a basic grasp of key individual terms (as shown by ROUGE-1 and ROUGE-L), its ability to replicate more complex structures or phrases from the reference text (as indicated by ROUGE-2) is quite limited. This suggests that while the model has some understanding of the domain-specific language, further fine-tuning is required for it to effectively replicate the more nuanced and structured aspects of the reference texts. Keep in mind that, as we have seen in other chapters, semantic similarity is also a good measure of domain-specific language understanding and does not rely on lexical overlap the way ROUGE does.

Qualitatively, domain experts should review the model's outputs to judge their relevance and accuracy in the context of Proxima's products and institutional language. These experts can provide insights into the nuances of the model's performance, which might not be captured by quantitative metrics alone. Comparing their feedback on the outputs from both the original and adapted models will highlight how well the adaptation has aligned BLOOM with Proxima's specific communication needs. This dual approach ensures a comprehensive evaluation, blending statistical analysis with real-world applicability and relevance.

Summary

In this chapter, we explored the domain adaptation process for the BLOOM LLM, which is specifically tailored to enhance its proficiency in the financial sector, particularly in understanding and generating content related to Proxima's product offerings. We began by introducing the concept of domain adaptation within the broader scope of transfer learning, emphasizing its significance in fine-tuning general-purpose models to grasp the intricacies of specialized fields.

The adaptation process involved integrating PEFT techniques into BLOOM and preprocessing a financial dataset for model training. This included standardizing text lengths through truncation and padding and tokenizing the texts for consistency in model input. The adapted model's performance was then quantitatively assessed against a reference dataset using the ROUGE metric, providing insights into its ability to capture key financial terminologies and phrases. Qualitative evaluation by domain experts was also suggested as a complementary method to gauge the model's practical effectiveness in real-world scenarios.

Overall, this chapter detailed a common approach to refining an LLM for a specific domain, illustrating both the methodology and the importance of a nuanced evaluation to ascertain the success of such adaptations. In the next chapter, we will explore how to adapt an LLM without fine-tuning using prompt engineering. We will discover how to contextualize and guide model outputs to produce similar results comparable to fine-tuned models.

References

This reference section serves as a repository of sources referenced within this book; you can explore these resources to further enhance your understanding and knowledge of the subject matter:

- Gururangan, S., Marasović, A., Swayamdipta, S., Lo, K., Beltagy, I., Downey, D., & Smith, N. A. (2020). *Don't stop pretraining: Adapt language models to domains and tasks*. In arXiv [cs.CL]. http://arxiv.org/abs/2004.10964/.

- Pruksachatkun, Y., Phang, J., Liu, H., Htut, P. M., Zhang, X., Pang, R. Y., Vania, C., Kann, K., & Bowman, S. R. (2020a). *Intermediate-task transfer learning with pretrained language models: When and why does it work?* Proceedings of the 58th Annual Meeting of the Association for Computational Linguistics.

- Xie, Q., Dai, Z., Hovy, E., Luong, M.-T., & Le, Q. V. (n.d.). *Unsupervised Data Augmentation for Consistency Training*. Arxiv.org. Retrieved March 16, 2024, from http://arxiv.org/abs/1904.12848.

- Anaby-Tavor, A., Carmeli, B., Goldbraich, E., Kantor, A., Kour, G., Shlomov, S., Tepper, N., & Zwerdling, N. (2020). *Do not have enough data? Deep learning to the rescue!* Proceedings of the ... AAAI Conference on Artificial Intelligence. AAAI Conference on Artificial Intelligence, 34(05), 7383–7390. https://doi.org/10.1609/aaai.v34i05.6233.

- Clark, K., Luong, M.-T., Khandelwal, U., Manning, C. D., & Le, Q. V. (2019). *BAM! Born-again multi-task networks for natural language understanding*. In arXiv [cs.CL]. http://arxiv.org/abs/1907.04829.

Mastering the Fundamentals of Prompt Engineering

In *Chapter 5*, we briefly evaluated a fine-tuned **Large Language Model** (**LLM**) against a general-purpose model using in-context learning or the few-shot prompting approach. In this chapter, we will revisit and explore prompting techniques to examine how well we can adapt a general-purpose LLM without fine-tuning. We explore various prompting strategies that leverage the model's inherent capabilities to produce targeted and contextually relevant outputs. We will start by examining the shift toward prompt-based language models. Then, we will revisit zero- and few-shot methods, explain prompt-chaining, and discuss various strategies, including more advanced techniques such as **Retrieval Augmented Generation** (**RAG**). At the end of the chapter, we will apply what we have learned and design a prompting strategy with the aim of consistently eliciting factual, accurate, and consistent responses that accomplish a specific business task.

Before diving into specific prompt engineering techniques, we will review a few breakthroughs that pioneered **State-of-the-Art** (**SOTA**) prompt-based models. Research from early 2018 demonstrated how pretraining LLMs could enable few-shot generalization – accurate performance on new tasks given only a prompt statement and a few demonstrations. Follow-up work further tailored model architectures and training specifically for excelling at prompt-based inference across many text-specific tasks. More recent methods optimized model efficiency and stability, enabling accurate and reliable and efficient prompt completion. These innovations laid the groundwork for prompt engineering, demonstrating the remarkable versatility of prompt-based models with minimal input data. Now, prompt design is becoming its own subfield of research – unlocking SOTA performance for an ever-expanding range of tasks. Let's get started.

The shift to prompt-based approaches

As discussed in prior chapters, the development of the original GPT marked a significant advance in natural language generation, introducing the use of prompts to instruct the model. This method allowed models such as GPT to perform tasks such as translations – converting text such as *"Hello, how are you?*" to "*Bonjour, comment ça va?*" – without task-specific training, leveraging deeply contextualized semantic patterns learned during pretraining. This concept of interacting with language models via natural language prompts was significantly expanded with OpenAI's GPT-3 in 2020. Unlike its predecessors, GPT-3 showcased remarkable capabilities in understanding and responding to prompts in zero- and few-shot learning scenarios, a stark contrast to earlier models that weren't as adept at such direct interactions. The methodologies, including the specific training strategies and datasets used for achieving GPT-3's advanced performance, remain largely undisclosed. Nonetheless, it is inferred from OpenAI's public research that the model learned to follow instructions based on its vast training corpus, and not explicit instruction-tuning. GPT-3's success in performing tasks based on simple and direct prompting highlighted the potential for language models to understand and execute a wide range of tasks without requiring explicit task-specific training data for each new task. This led to a new paradigm in NLP research and applications, focusing on how effectively a model could be prompted with instructions to perform tasks such as summarization, translation, content generation, and more.

After the release of GPT-3, OpenAI was among the first to introduce specialized fine-tuning to respond more accurately to instructions in their release of InstructGPT (Ouyang et al., 2022). The researchers aimed to teach the model to closely follow instructions using two novel approaches. The first was **Supervised Fine-Tuning (SFT)**, which involved fine-tuning using datasets carefully crafted from prompts and response pairs. These *demonstration* datasets were then used to perform SFT on top of the GPT-3 pretrained model, refining it to provide responses more closely aligned with human responses. *Figure 7.1* provides an example of a prompt and response pair.

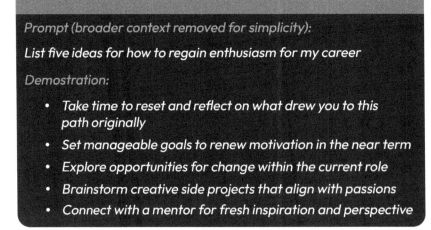

Figure 7.1: InstructGPT SFT instruction and output pairs

The second approach involved additional refinement using **Reinforcement Learning from Human Feedback (RLHF)**. **Reinforcement Learning (RL)**, established decades ago, aims to enhance autonomous agents' decision-making capabilities. It does this by teaching them to optimize their actions based on the trade-off between risk and reward. The policy captures the guidelines for the agent's behavior, dynamically updating as new insights and feedback are learned to refine decisions further. RL is the exact technology used in many robotic applications and is most famously applied to autonomous driving.

RLHF is a variation of traditional RL, incorporating human feedback alongside the usual risk/reward signals to direct LLM behavior toward better alignment with human judgment. In practice, human labelers would provide preference ratings on model outputs from various prompts, and these ratings would be used to update the model policy, steering the LLM to generate responses that better conform to expected user intent across a range of tasks. In effect, this technique helped to reduce the model's tendency to generate inappropriate, biased, harmful, or otherwise undesirable content. Although RLHF is not a perfect solution in this regard, it represents a significant step toward models that better understand and align with human values.

Later that year, following OpenAI's introduction of InstructGPT, Google unveiled **Fine-tuned Language Net** or **FLAN** (Wei et al., 2021). FLAN represented another leap toward prompt-based LLMs, employing explicit instruction tuning. Google's approach relied on formatting existing datasets into instructions, enabling the model to understand various tasks. Specifically, the authors of FLAN merged multiple NLP datasets across different categories, such as translation and question answering, creating distinct instruction templates for each dataset to frame them as instruction-following tasks. For example, the FLAN team leveraged ANLI challenges (Nie et al., 2020) to construct question-answer pairs explicitly designed to test the model's understanding of complex textual relationships and reasoning. By framing these challenges as question-answer pairs, the FLAN team could directly measure a model's proficiency in deducing these relationships under a unified instruction-following framework. Through this innovative approach, FLAN effectively broadened the scope of tasks a model can learn from, enhancing its overall performance and adaptability across a diverse set of NLU benchmarks. *Figure 7.2* presents a theoretical example of question-answer pairs based on ANLI.

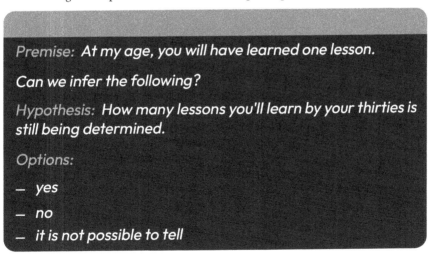

Figure 7.2: Training templates based on the ANLI dataset

Again, the central idea behind FLAN was that each benchmark dataset (e.g., ANLI) could be translated into an intuitive instruction format, yielding a broad mixture of instructional data and natural language tasks.

These advancements, among others, represent a significant evolution in the capabilities of LLMs, transitioning from models that required specific training for each task to those that can intuitively follow instructions and adapt to a multitude of tasks with a simple prompt. This shift has not only broadened the scope of tasks these models can perform but also demonstrated the potential for AI to process and generate human language in complex ways with unprecedented precision.

With this insight, we can shift our focus to prompt engineering. This discipline combines technical skill, creativity, and human psychology to maximize how models comprehend and respond, appropriately and accurately, to instructions. We will learn prompting techniques that increasingly influence the model's behavior toward precision.

Basic prompting – guiding principles, types, and structures

In *Chapter 5*, we introduced the concept of zero- and few-shot learning, providing the model either a direct instruction, or a direct instruction paired with examples specific to the task. In this section, we will focus on zero-shot learning, where prompting becomes a critical tool for guiding the model to perform specific tasks without prior explicit training on those tasks. This section explores elements of a prompt and how to structure it effectively for zero-shot learning. However, we will first establish some critical guiding principles to help us understand expected model behavior.

Guiding principles for model interaction

It is absolutely critical to understand that LLMs, despite their unprecedented SOTA performance on natural language tasks, have significant inherent limitations, weaknesses, and susceptibilities. As described in *Chapter 1*, LLMs cannot establish rationale or perform logical operations natively. Our interactions with LLMs are typically supplemented by a highly sophisticated application layer that enables the raw model to carry on an extended exchange, integrate with systems that perform computations, and retrieve additional information and knowledge not intrinsic to the model itself. Independent of supplemental integrations, many LLMs are prone to erratic behavior. The most common of these is often referred to as **hallucination**, where the model generates a plausible output that is not entirely factual. As such, we should approach the general use of LLMs with the following guidelines in mind:

- **Apply domain knowledge and subject-matter expertise**: As SOTA LLMs are prone to generating inaccuracies that sound plausible, in use cases where factuality and precision are essential (e.g., code generation, technical writing, or academic research), users must have a firm grasp of the subject matter to detect potential inaccuracies. For example, suppose a user without medical expertise were to prompt a model for healthcare advice. In that case, the model may confuse, conflate, or simply invent information that could result in misleading or potentially dangerous advice. A mitigant for this behavior could be to provide the model with information from a reputable health journal and instruct it to generate its answers explicitly from the passages provided. This technique is often called grounding, and we will cover it in depth later. However, even when supplementing the model's knowledge with verified information, the model can still misrepresent facts. Without expertise in the specific domain in question, we may never detect misinformation. Consequently, we should generally avoid using LLMs when we cannot verify the model output. Moreover, we should avoid using LLMs in high-stake scenarios where erroneous output could have profound implications.

- **Acknowledge bias, underrepresentation, and toxicity**: We have described how LLMs are trained at an enormous scale and often on uncurated datasets. Inevitably, LLMs will learn, exhibit, and amplify societal biases. The model will propagate stereotypes, reflect biased assumptions, and generate toxic and harmful content. Moreover, LLMs can overrepresent certain populations and grossly underrepresent others, leading to a skewed or warped sociological perspective. These notions of bias can manifest in many ways. We will explore this topic, and other ethical implications of LLM use, in detail in *Chapter 8*.

- **Avoid ambiguity and lack of clarity**: Since LLMs were trained to synthesize information resembling human responses, they can often exhibit notions of creativity. In practice, if prompting is ambiguous or lacks clarity, the model will likely use its vast contextualized knowledge to "assume" or "infer" the meaning or objective of a given prompt or instruction. It may apply some context from its training instead of responding with a clarifying question. As we will describe in the next section, it is crucial to provide clarity by contextualizing input in most cases.

Now that we have established a few overarching principles to help navigate interactions and keep us within the boundaries of appropriate use, we can deconstruct the various elements of a prompt.

Prompt elements and structure

Generally, a prompt acts as a guide, directing the model's response toward the desired outcome. It typically comprises key elements that frame the task at hand, providing clarity and direction for the model's generative capabilities. The following table presents the essential elements of a zero-shot prompt.

Instruction	A clear, concise statement describing what you want the model to do. This could be a direct command, a question, or a statement that implies a task.
Context	Relevant information or background is needed to understand the instruction or the task. This could include definitions or clarifications.
Input	Following the instructions, the model should work with specific data or content. This could be a piece of text, a question, or any information relevant to the task.
Output cue	An indication of how the model's response is to be structured. This can be part of the instruction or implied through the prompt's formatting.

Table 7.1: Basic elements of a zero-shot prompt

We can then structure these elements to maximize the zero-shot approach, whereby the model relies entirely on the prompt to understand and execute a task. In this context, we use the term *task* to describe a specific natural language task, such as summarization or translation. However, we will also encounter the term *task* applied more broadly to refer to the output the model should provide. Let's explore a few concrete examples of various tasks. In this case, we will be referring to specific NLP tasks and applying a standard structure combining the key elements we've described:

- **Example 1: Summarization task**

  Instruction: Summarize the following text in one sentence.

  Context: The text provides an overview of the benefits of renewable energy.

  Input: `Renewable energy sources like solar and wind power offer sustainable alternatives to fossil fuels, reducing greenhouse gas emissions and promoting environmental conservation...`

  Output Cue: Renewable energy sources, such as

  Example Outcome: `"Renewable energy sources, such as solar and wind, play a crucial role in reducing emissions and conserving the environment."`

- **Example 2: Translation task**

  Instruction: Translate the following sentence from English to Spanish.

  Context: The sentence is a greeting.

  Input: `"Hello, how are you?"`

  Output Cue: This translates to

  Example Outcome: `This translates to "Hola, ¿cómo estás?"`

  The structured templates help us to efficiently and reliably prompt the model for a wide range of inputs, while maintaining a structure that the model has learned to recognize and respond to. In fact, we can take this a step further by asking the model to provide a specific format in its output. Using the output cue, we can instruct the model to provide a specified format such as Markdown.

- **Example 3: Code generation task**

  Instruction: Generate a Python function that calculates the square of a number.

  Context: The function should take a single integer argument and return its square.

  Input: `"Please write a Python function to calculate the square of a number."`

Output Cue: By using the Markdown format in the output cue, the model knows to provide this format and returns the following:

```
def square(number):
    return number ** 2
```

Using LangChain to produce JSON-formatted output, we can leverage the same approach. Specifically, LangChain's `PromptTemplate` provides a flexible way to dynamically define a structure for our prompts and insert elements:

```
from langchain.prompts import PromptTemplate
from langchain.llms import OpenAI

# Define a prompt template requesting JSON formatted output
prompt_structure = PromptTemplate(
    template="""
        Context: {context}
        Instruction: {instruction}
        Text: {text_to_process}
        Output Cue: Format the response in JSON with one element
called summary.
    """,
    input_variables=["context," "instruction",
        "text_to_process"]
)


# Dynamic elements for the prompt
context = "Summarizing long text passages."
instruction = "Summarize the key points from the following text
in JSON format."
text_to_process = """
Mars is the fourth planet from the Sun. The surface of Mars is
orange-red because…
"""


formatted_prompt = prompt_structure.format_prompt(
    context=context,
    instruction=instruction,
    text_to_process=text_to_process
)


llm = OpenAI(model_name='gpt-3.5-turbo-instruct',
    temperature=0.9, max_tokens = 256)
response = llm.invoke(formatted_prompt)
print(response)
```

This produces the following:

```
{
     "summary": "Mars is the fourth planet from the Sun, known
for its orange-red surface and high-contrast features that make
it a popular object for telescope viewing."
}
```

Crafting effective prompts for zero-shot learning with LLMs requires a clear understanding of the task, thoughtful structuring of the prompt, and consideration of how the model interprets and responds to different elements within the prompt. By applying these principles, we can guide models to perform various tasks accurately and effectively. Subsequently, we will explore methods to guide models' behavior through positive affirmations, emotional engagement, and other cognitive-behavioral techniques.

Elevating prompts – iteration and influencing model behaviors

In this section, we will introduce techniques for enhancing AI model interactions inspired by cognitive-behavioral research. Behavioral prompting can guide models toward more accurate and nuanced responses. For example, LLM performance can be improved by providing the model with positive emotional stimuli, asking the model to assume a persona or character, or using situational prompting (i.e., role-play). However, it is crucial to recognize that these techniques can also be misused or used to inadvertently introduce stereotypes, as they rely on assumptions and generalizations that may not accurately reflect individual experiences or diverse perspectives. Without careful consideration and monitoring, there is a risk of reinforcing existing biases or creating new ones, potentially leading to skewed or harmful output. Given these challenges, we will explore a responsible approach to employing cognitive-behavioral techniques in AI interactions, aiming to harness their benefits while minimizing risks and ensuring inclusivity and fairness.

LLMs respond to emotional cues

Research conducted by Microsoft in collaboration with various institutions, including the Beijing Normal University psychology department, suggests that LLMs can mimic and display some aspects of human emotional intelligence. This can lead to improved task performance when prompts are infused with emotional stimuli. In particular, the researchers hypothesize that emphasizing positive words can trigger more constructive and effective responses. The phenomenon is not well understood, but the effect is that positive emotional cues seem to improve model performance on various tasks consistently (Li et al., 2023). For example, the researchers input phrases encouraging confidence and positive outcomes, such as This is very important to my career or Believe in your abilities and strive for excellence. Your hard work will yield remarkable results. These types of psychological elements not only positively influenced the output quality but also made interactions more effective and nuanced.

Effect of personas

Research has shown that the effect of **personas** enables LLMs to better simulate human-like interactions by adopting varied characteristics, perspectives, and response styles tailored to specific user profiles. This approach enhances the model's capability for engaging in more personalized, contextually relevant conversations. Incorporating personas into LLM prompts effectively shifts the distribution of possible outputs, directing the model's generation process toward responses that resonate with specific character traits or perspectives. Thanks to the model's extensive training on diverse datasets, it can replicate patterns aligning with the described characteristics of personas, focusing its attention and selection process on a more defined set of outputs. This enhances relevance and specificity.

For clarity, let's reconsider our hypothetical business StyleSprint, a fashion brand. If we prompt an LLM with "`As a creative and bold fashion brand, how would StyleSprint describe its upcoming summer collection?`", the model, adopting the brand's persona, might generate a response such as the following:

"`This summer, StyleSprint dares you to stand out! Our collection is a vibrant explosion of colors and patterns, designed for the fearless fashionista. Think bold prints, daring cuts, and a statement that screams confidence. From sun-drenched beaches to chic city streets, our pieces promise to make every moment unforgettable. Embrace your boldness with StyleSprint this summer.`".

In this example, we can see how personas guide the model's outputs to reflect specific brand tones, such as creativity and boldness in this case, making the marketing material more engaging and aligned with the brand's identity. Furthermore, using a few-shot approach, we could imbue model responses with a very specific branded tone. We will discuss few-shot in the sections that follow.

However, as discussed, personas should be used with caution. Personas can perpetuate stereotypes and biases, particularly against marginalized groups. A study conducted by researchers at Stanford University found that generating personas based on intersectional demographic groups often yields higher rates of racial stereotypes and patterns of **othering,** or portraying someone or a group as fundamentally different or alien, compared to human-written texts. In some cases, model outputs could amplify narratives and tropes (Cheng, Durmus, & Jurafsky, 2023).

Situational prompting or role-play

Role-play in LLMs, in the same way as personas, involves adopting specific identities or characteristics. However, the two serve different purposes and are applied in distinct contexts. Personas are predefined sets of traits or characteristics that an LLM mimics to tailor its responses, focusing on consistency with those traits. As we have demonstrated with our StyleSprint example, this is useful for creating content with a specific tone or perspective.

Conversely, role-play extends beyond adopting a set of traits to engage in a scenario or narrative dynamically. It involves the LLM taking on a character within a simulated environment or story, responding to inputs in a manner that aligns with both a persona and the evolving context of the role-play scenario. This can be especially useful in complex simulations where the LLM must navigate and contribute to ongoing narratives or dialogues that require understanding and adapting to new information or changing circumstances in real time.

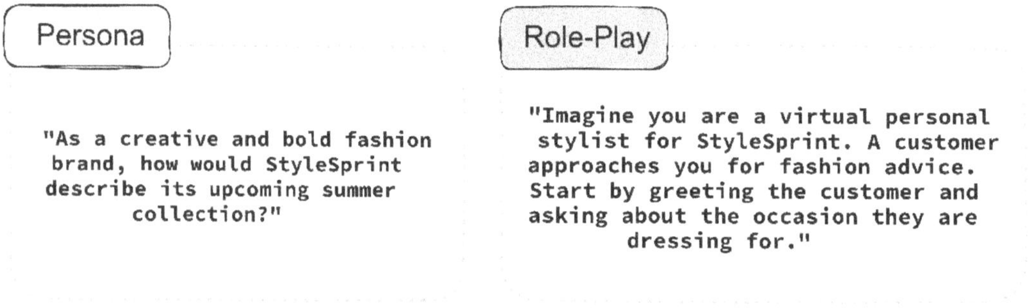

Figure 7.3: Persona versus role-play

Revisiting our real-world scenario, role-play could be particularly useful for creating interactive and engaging customer service experiences. For example, StyleSprint could design a role-play scenario where the LLM acts as a virtual personal stylist. In this role, the model would engage customers with prompts such as `I'm your personal stylist for today! What's the occasion you're dressing for?`. Based on the customer's response, the LLM could ask follow-up questions to narrow down preferences, such as `Do you prefer bold colors or pastel shades?`. Finally, it could recommend outfits from StyleSprint's collection that match the customer's needs, saying something such as `For a summer wedding, I recommend our Floral Maxi Dress paired with the Vintage Sun Hat. It's elegant, yet perfect for an outdoor setting!`.

In this case, we leverage the LLM's ability to dynamically adapt its dialogue based on customer inputs to create an advanced recommender system that facilitates a highly personalized shopping experience. It not only helps in providing tailored fashion advice but also engages customers in a novel way.

Having examined how behavior-inspired techniques, such as personas and role-play, influence model behavior through zero-shot learning, let's now turn our attention to few-shot learning. This is also known as in-context learning, which we described in *Chapter 5*. Recall that the few-shot approach can enhance the consistency, stability, and reliability of model responses. By providing the model with a few examples of the desired output within the prompt itself, few-shot learning effectively teaches the model the specific task at hand, leading to more predictable and accurate outputs.

Advanced prompting in action – few-shot learning and prompt chaining

In few-shot settings, the LLM is presented with a small number of examples of a task within the input prompt, guiding the model to generate responses that align with these examples. As discussed in the prior chapter, this method significantly reduces the need for fine-tuning on large, task-specific datasets. Instead, it leverages the model's pre-existing knowledge and ability to infer context from the examples provided. In *Chapter 5*, we saw how this approach was particularly useful for StyleSprint by enabling the model to answer specific questions after being provided with just a few examples, enhancing consistency and creativity in brand messaging.

This method typically involves using between 10 and 100 examples, depending on the model's context window. Recall that the context window is the limit of tokens a language model can process in one turn. The primary benefit of the few-shot approach is that it minimizes the risk of the model learning a too-narrow distribution from a specific dataset through fine-tuning. Although the performance of few-shot may not always match its fine-tuned counterpart, few-shot learning often outperforms both one-shot and zero-shot learning, showing significant improvements in task adaptation and accuracy. This is especially true as more examples are added to the context window (Brown et al., 2020).

Applications such as LangChain provide a simple and convenient pattern for few-shot implementation. Consider a scenario in which StyleSprint would like to generate taglines for its seasonal collections. In this case, we can provide the model with examples written by the content team to guide the model toward consistency with the brand tone:

```
examples = [
    {
        "prompt": "Describe the new summer collection in a bold and
adventurous tone.",
        "response": "Dive into summer with StyleSprint's latest
collection! Featuring daring designs and vibrant colors, it's all
about making bold statements. Perfect for the fearless fashionista
ready to conquer the heat."
    },
    {
        "prompt": "How would you introduce our eco-friendly line to
environmentally conscious customers?",
        "response": "Embrace sustainable style with StyleSprint's eco-
friendly line. Crafted from recycled materials, each piece combines
fashion with responsibility, designed for the eco-conscious and
trendy."
    }
]
```

The LangChain API offers `FewShotPromptTemplate` to format the examples consistently:

```
from langchain.prompts.few_shot import FewShotPromptTemplate
from langchain.prompts.prompt import PromptTemplate

# Create a formatter
prompt_format = PromptTemplate(
    input_variables=["prompt", "response"],
    template="Prompt: {prompt}\nResponse: {response}")

# Create the FewShotPromptTemplate
few_shot_prompt = FewShotPromptTemplate(
    examples=examples, example_prompt=prompt_format,
    suffix="Prompt: {input}", input_variables=["input"])
```

We can now apply the template to an LLM to generate a response that we can expect will closely align with the tone and style of our examples:

```
from langchain import LLMChain, OpenAI

# Setup the LLM and LLMChain
llm = OpenAI(temperature=0)
llm_chain = LLMChain(llm=llm, prompt=few_shot_prompt)

# Define the input prompt
input_prompt = "Create a catchy tagline for our winter collection."

# Invoke the chain to generate output
response = llm_chain.run(input_prompt)

# Extract and print the generated slogan
generated_slogan = response
print(generated_slogan)
    # => Response: "Stay warm,
    stay stylish,
    stay ahead with StyleSprint's winter collection!"
```

Now that we have a consistent and programmatic method for providing the model with examples, we can iterate over the model responses using prompt chaining. A prompt chain generally refers to chaining together multiple prompts and LLM interactions to have a conversation with the model and iteratively build on the results. Remember, the model itself cannot store information and effectively has no memory or prior inputs and outputs. Instead, the application layer stores prior inputs and outputs, which are then provided to the model with each exchange. For example, you might start with an initial prompt such as the following:

```
"Write a slogan for a winter clothing line"
```

The LLM might generate the following:

```
"Be warm, be cozy, be you"
```

You could then construct a follow-up prompt using the following:

```
"Modify the slogan to be more specific about the quality of the
clothing"
```

You could then keep iterating to improve the output.

Chaining facilitates guiding and interactively refining the text generated rather than relying purely on the given examples. Notice that our prior few-shot code had already established a chain, which we can now use to iterate as follows:

```
response = llm_chain.run("Rewrite the last tag to something about
embracing the winter")
Response #
=> Response: Embrace the winter wonderland with StyleSprint's latest
collection. From cozy knits to chic outerwear, our pieces will keep
you stylish and warm all season long.
```

The model is now working from both the examples we provided and any additional instructions we want to include as part of the chain. Prompt chaining, combined with few-shot learning, provides a powerful framework for iteratively guiding language model outputs. By leveraging application state to maintain conversation context, we can steer the model toward desired responses in line with our provided examples. This approach balances harnessing the model's inferential capabilities and retaining control to align its creative outputs.

Next, we will dive into our practice project, which implements RAG. RAG augments model responses by retrieving and incorporating external data sources. This technique mitigates hallucination risks by grounding AI-generated text in real data. For example, StyleSprint may leverage past customer survey results or catalog data to enhance product descriptions. By combining retrieval with prompt chaining, RAG provides a scalable method for balancing creativity with accuracy.

Practice project: Implementing RAG with LlamaIndex using Python

For our practice project, we will shift from LangChain to exploring another library that facilitates the RAG approach. LlamaIndex is an open source library that is specifically designed for RAG-based applications. LlamaIndex simplifies ingestion and indexing across various data sources. However, before we dive into implementation, we will explain the underlying methods and approach behind RAG.

As discussed, the key premise of RAG is to enhance LLM outputs by supplying relevant context from external data sources. These sources should provide specific and verified information to ground model outputs. Moreover, RAG can optionally leverage the few-shot approach by retrieving few-shot examples at inference time to guide generation. This approach alleviates the need to store examples in the prompt chain and only retrieves relevant examples when needed. In essence, the RAG approach is a culmination of many of the prompt engineering techniques we have already discussed. It provides structure, chaining, few-shot learning, and grounding.

At a high level, the RAG pipeline can be described as follows:

1. The RAG component ingests and indexes domain-specific data sources using vector embeddings to encode semantics. As we learned in *Chapter 3*, these embeddings are imbued with deeply contextualized, rich semantic information that the component uses later to perform a semantic search.

2. The component then uses the initial prompt as a search query. The query is input to retrieval systems, which find the most relevant snippets from the indexed data based on vector similarity. Similar to how we applied semantic similarity in prior chapters, RAG leverages a similarity metric to rank results by semantic relevance.

3. Lastly, the original prompt is augmented with information from the retrieved contexts, and the augmented prompt is passed to the LLM to generate a response grounded in the external data.

RAG introduces two major benefits. First, like the chaining approach, the indexed external data acts as a form of memory, overcoming the LLM's statelessness. Second, this memory can rapidly scale beyond model context window limitations, since examples are curated and only provided at the time of the request as needed. Ultimately, RAG unlocks otherwise unattainable capabilities in reliable and factual text generation.

In our practice project, we revisited the StyleSprint product descriptions. This time, we want to leverage RAG to retrieve detailed information about the product to produce very specific descriptions. For the purpose of keeping this project accessible, we will implement an in-memory vector store (Faiss) instead of an external database. We begin with installing the necessary libraries. We will leverage LlamaIndex's integrated support for Faiss:

```
pip install llama-index faiss-cpu llama-index-vector-stores-faiss
```

We will then import the necessary libraries, load the data, and create the index. This vector store will rely on OpenAI's embeddings, so we must also define OPENAI_API_KEY using a valid key:

```
assert os.getenv("OPENAI_API_KEY") is not None,
    "Please set OPENAI_API_KEY"

# load document vectors
documents = SimpleDirectoryReader("products/").load_data()

# load faiss index
d = 1536 # dimension of the vectors
faiss_index = faiss.IndexFlatL2(d)
```

```
# create vector store
vector_store = FaissVectorStore(faiss_index=faiss_index)

# initialize storage context
storage_context = StorageContext.from_defaults(
    vector_store=vector_store)

# create index
index = VectorStoreIndex.from_documents(
    documents,storage_context=storage_context)
```

We now have a vector store that the model can rely on to retrieve our very specific product data. This means we can query for very specific responses augmented by our data:

```
# query the index
query_engine = index.as_query_engine()
response = query_engine.query("describe summer dress with price")

print(response)
=> A lightweight summer dress with a vibrant floral print is priced at
59.99.
```

The result is a response that not only provides an accurate description of the summer dress but also includes specific details, such as the price. This level of detail enriches the customer's shopping experience, providing relevant and real-time information for customers to consider when making a purchase.

The next step is to evaluate our RAG implementation to ensure that the answer is relevant, faithful to the source text, reflective of contextual accuracy, and not in any way harmful or inappropriate. We can apply an open source evaluation framework (RAGAS), which provides implementation of the following metrics:

- **Faithfulness** assesses the degree to which the generated response is faithful or true to the original context

- **Answer relevance** evaluates how relevant the generated answer is to the given question

- **Context precision** measures the precision of the context used to generate the answer

- **Context recall** measures the recall of the context used to generate the answer

- **Context relevancy** assesses the relevancy of the context used to generate the answer

- **Harmfulness** evaluates whether a submission (or answer) contains anything that could potentially cause harm to individuals, groups, or society at large

This suite of metrics provides an objective measure of RAG application performance based on a comparison to ground truth. In our case, we can use responses generated from our product data, along with context and ground truth derived from the original dataset, to construct an evaluation dataset and perform a comprehensive evaluation using the metrics described.

The following is a simplified code snippet implementing the RAGAS evaluation for our generated product descriptions. A complete working implementation is available in the Chapter 7 folder of the GitHub companion to this book (https://github.com/PacktPublishing/Generative-AI-Foundations-in-Python).

```
# Define the evaluation data
eval_data: Dict[str, Any] = {
    "question": questions, # list of sampled questions
    "answer": engine_responses, # responses from RAG application
    "contexts": contexts, # product metadata
"ground_truth": ground_truth, # corresponding descriptions written by
a human
}


# Create a dataset from the evaluation data
dataset: Dataset = Dataset.from_dict(eval_data)


# Define the evaluation metrics
metrics: List[Callable] = [
    faithfulness,
    answer_relevancy,
    context_precision,
    context_recall,
    context_relevancy,
    harmfulness,
]
# Evaluate the model using the defined metrics
result: Dict[str, float] = evaluate(dataset, metrics=metrics)
print(result)
```

Our evaluation program should produce the following:

```
{'faithfulness': 0.9167, 'answer_relevancy': 0.9961, 'context_
precision': 0.5000, 'context_recall': 0.7500, 'harmfulness': 0.0000}
```

We can observe that the system performs well in generating accurate and relevant answers, as evidenced by high faithfulness and answer relevancy scores. While context precision shows room for improvement, half of the relevant information is correctly identified. Context recall is effective, retrieving most of the relevant context. The absence of harmful content ensures safe interactions. Overall, the system displays robust performance in answering accurately and contextually, but could benefit from refinements in pinpointing the most pertinent context snippets.

As discussed in *Chapters 5* and *6*, the evaluation of LLMs often requires the additional operational burden of collecting ground-truth data. However, doing so makes it possible to perform a robust evaluation of model and application performance.

Summary

In this chapter, we explored the intricacies of prompt engineering. We also explored advanced strategies to elicit precise and consistent responses from LLMs, offering a versatile alternative to fine-tuning. We traced the evolution of instruction-based models, highlighting how they've shifted the paradigm toward an intuitive understanding and adaptation to tasks through simple prompts. We expanded on the adaptability of LLMs with techniques such as few-shot learning and retrieval augmentation, which allow for dynamic model guidance across diverse tasks with minimal explicit training. The chapter further explored the structuring of effective prompts, and the use of personas and situational prompting to tailor model responses more closely to specific interaction contexts, enhancing the model's applicability and interaction quality. We also addressed the nuanced aspects of prompt engineering, including the influence of emotional cues on model performance and the implementation of RLHF to refine model outputs. These discussions underscored the potential of LLMs to exhibit some level of emotional intelligence, leading to more effective and nuanced interactions. However, alongside these technological strides, we stressed the paramount importance of ethical considerations. We highlighted the need for responsible adoption and vigilance to mitigate potential harm and biases associated with these techniques, ensuring fairness, integrity, and the prevention of misuse.

Lastly, we learned how to implement and evaluate the RAG approach to ground the LLM in contextual information from trusted sources and produce answers that are relevant and faithful to the source text. In the next chapter, we will look more closely at the role of individuals in advancing generative AI while emphasizing the dual responsibility of developers and researchers to navigate this rapidly evolving field with a conscientious approach, balancing innovation with ethical imperatives and societal impacts.

References

This reference section serves as a repository of sources referenced within this book; you can explore these resources to further enhance your understanding and knowledge of the subject matter:

- Ouyang, L., Wu, J., Jiang, X., Almeida, D., Wainwright, C. L., Mishkin, P., Zhang, C., Agarwal, S., Slama, K., Ray, A., Schulman, J., Hilton, J., Kelton, F., Miller, L., Simens, M., Askell, A., Welinder, P., Christiano, P., Leike, J., & Lowe, R. (2022). *Training language models to follow instructions with human feedback*. In arXiv [cs.CL]. `http://arxiv.org/abs/2203.02155`

- Wei, J., Bosma, M., Zhao, V. Y., Guu, K., Yu, A. W., Lester, B., Du, N., Dai, A. M., & Le, Q. V. (2021). *Finetuned language models are zero-shot learners*. In arXiv [cs.CL]. `http://arxiv.org/abs/2109.01652`

- Nie, Y., Williams, A., Dinan, E., Bansal, M., Weston, J., & Kiela, D. (2020). *Adversarial NLI: A new benchmark for natural language understanding*. Arxiv.org.

- Li, C., Wang, J., Zhang, Y., Zhu, K., Hou, W., Lian, J., Luo, F., Yang, Q., & Xie, X. (2023). *Large Language Models understand and can be enhanced by emotional stimuli*. In arXiv [cs.CL]. `http://arxiv.org/abs/2307.11760`

- Cheng, M., Durmus, E., & Jurafsky, D. (2023). *Marked personas: Using natural language prompts to measure stereotypes in language models*. Proceedings of the 61st Annual Meeting of the Association for Computational Linguistics (Volume 1: Long Papers).

- Brown, T. B., Mann, B., Ryder, N., Subbiah, M., Kaplan, J., Dhariwal, P., Neelakantan, A., Shyam, P., Sastry, G., Askell, A., Agarwal, S., Herbert-Voss, A., Krueger, G., Henighan, T., Child, R., Ramesh, A., Ziegler, D. M., Wu, J., Winter, C., ... Amodei, D. (2020). *Language Models are Few-Shot Learners*. In arXiv [cs.CL]. `http://arxiv.org/abs/2005.14165`

## 8

Addressing Ethical Considerations and Charting a Path Toward Trustworthy Generative AI

As generative AI advances, it will extend beyond basic language tasks, integrating into daily life and impacting almost every sector. The inevitability of its widespread adoption highlights the need to address its ethical implications. The promise of this technology to revolutionize industries, enhance creativity, and solve complex problems must be coupled with the responsibility to navigate its ethical landscape diligently. This chapter will explore these ethical considerations, dissect the intricacies of biases entangled in these models, and look at strategies for cultivating trust in general-purpose AI systems. Through thorough examination and reflection, we can begin to outline a path toward responsible use, helping to ensure that advancements in generative AI are leveraged for the greater good while minimizing harm.

To ground our discussion, we will first identify some ethical norms and universal values relevant to generative AI. While this chapter cannot be exhaustive, it aims to introduce key ethical considerations.

Ethical norms and values in the context of generative AI

The ethical norms and values guiding the development and deployment of generative AI are rooted in transparency, equity, accountability, privacy, consent, security, and inclusivity. These principles can serve as a foundation for developing and adopting systems aligned with societal values and supporting the greater good. Let's explore these in detail:

- **Transparency** involves clearly explaining the methodologies, data sources, and processes behind large language model (LLM) construction. This practice builds trust by enabling stakeholders to understand the technology's reliability and limits. For example, a company could publish a detailed report on the types of data trained on their LLM and the steps taken to ensure data privacy and bias mitigation.

- **Equity** in the context of LLMs ensures fair treatment and outcomes for all users by actively preventing biases in models. This requires thorough analysis and correction of training data and continuous monitoring of exchanges to reduce discrimination. One measure a firm might apply is a routine review of LLM performance across various demographic groups to identify and address unintended biases.

- **Accountability** establishes that developers and users of LLMs are responsible for model outputs and impacts. It includes transparent and accessible mechanisms for reporting and addressing negative consequences or ethical violations. In practice, this could manifest as the establishment of an independent review board that oversees AI projects and intervenes in cases of ethical misconduct.

- **Privacy and consent**, in principle, involves ensuring that individual privacy and consent are respected and preserved during the use of personal data as input to LLMs. In practice, developers should avoid using personal data for training without explicit permission and implement strong data protection measures. For example, a developer might use data anonymization or privacy-preserving techniques to train models, ensuring that personal identifiers and sensitive information are removed before data processing.

- **Security** involves protecting LLM-integrated systems and their data from unauthorized access and cyber threats. In practice, setting up LLM-specific red teams (or teams that test defenses by simulating attacks) can help safeguard AI systems against potential breaches.

- **Inclusivity** involves the deliberate effort to include diverse voices and perspectives in the development process of LLMs, ensuring the technology is accessible and beneficial to a broad spectrum of users. In practice, it is vital to collaborate with socio-technical subject-matter experts who can guide appropriate actions to promote and preserve inclusion.

This set of principles is not comprehensive but may help to form a conceptual foundation for ethical LLM development and adoption with the universal goal of advancing the technology in ways that avoid harm.

Additionally, various leading authorities have published guidance regarding responsible AI, inclusive of ethical implications. These include the US Department of Commerce's **National Institute of Standards and Technology (NIST)**, Stanford University's **Institute for Human-Centered Artificial Intelligence (HAI)**, and the **Distributed AI Research Institute (DAIR)**, to name a few.

Investigating and minimizing bias in generative LLMs and generative image models

Bias in generative AI models, including both LLMs and generative image models, is a complex issue that requires careful investigation and mitigation strategies. Bias can manifest as unintended stereotypes, inaccuracies, and exclusions in the generated outputs, often stemming from biased datasets and model architectures. Recognizing and addressing these biases is crucial to creating equitable and trustworthy AI systems.

At its core, algorithmic or model bias refers to systematic errors that lead to preferential treatment or unfair outcomes for certain groups. In generative AI, this can appear as gender, racial, or socioeconomic biases in outputs, often mirroring societal stereotypes. For example, an LLM may produce content that reinforces these biases, reflecting the historical and societal biases present in its training data.

Let us again revisit our hypothetical fashion retailer, StyleSprint. Consider a situation where StyleSprint experimented with using a multimodal generative LLM model to generate promotional images and captions for its latest sneaker line. It finds that the model predominantly generates sneakers in urban, graffiti-laden backgrounds, unintentionally drawing an association that relies on stereotypes. Moreover, the team begins noticing that the captions are also laden with language that perpetuates stereotypes. This realization prompts a reevaluation of the imagery and text, first with an investigation of how the problem surfaced.

Investigating bias involves various techniques, from analyzing the diversity and representativeness of training datasets to implementing testing protocols that specifically look for biased outputs across different demographics and scenarios. Statistical analysis can reveal disparities in model outcomes, while comparative studies and user feedback can help identify biases in the generated content.

In this case, let us assume that StyleSprint was using an LLM-provider without the ability to influence its training data or development process. To mitigate the risk of bias, the team might employ the following:

- Post-processing adjustments to diversify the imagery, ensuring a broader representation of backgrounds that resonate with its customer base

- The institution of a manual review process, enlisting team members to scrutinize and curate AI-generated images and captions before publishing (i.e., "human-in-the-loop"), ensuring that every piece of content aligns with the brand's commitment to diversity and inclusion

As is true for other kinds of evaluation of generative AI, evaluating bias demands both quantitative and qualitative methods. Statistical analysis can uncover performance disparities across groups, and comparative studies can detect biases in outputs. Gathering feedback from diverse users aids the understanding of real-world bias impacts, while independent audits and research are essential for identifying issues that internal evaluations may miss.

With a better understanding of how we might investigate and evaluate model outcomes for societal bias, we can explore technical methods for guiding model outcomes toward reliability, equity, and general trustworthiness to curb biased or inequitable outcomes during inference.

Constrained generation and eliciting trustworthy outcomes

In practice, it is possible to constrain model generation and guide outcomes toward factuality and equitable outcomes. As discussed, guiding models toward trustworthy outcomes can be done through continued training and fine-tuning, or during inference. For example, methodologies such as **reinforcement learning from human feedback** (RLHF) and **direct preference optimization** (DPO) increasingly refine model outputs to align model outcomes with human judgment. Additionally, as discussed in *Chapter 7*, various grounding techniques help to ensure that model outputs reflect verified data, continuously guiding the model toward responsible and accurate content generation.

Constrained generation with fine-tuning

Refinement strategies such as RLHF integrate human judgments into the model training process, steering the AI toward behavior that aligns with ethical and truthful standards. By incorporating human feedback loops, RLHF ensures that the AI's outputs meet technical accuracy and societal norms.

Similarly, DPO refines model outputs based on explicit human preferences, providing precise control to ensure outcomes adhere to ethical standards and human values. This technique exemplifies the shift toward more ethically aligned content generation by directly incorporating human values into the optimization process.

Constrained generation through prompt engineering

As we discovered in *Chapter 7*, we can guide model responses by grounding the LLM with factual information. This can be achieved directly using the context window or retrieval approach (e.g., Retrieval Augmented Generation (RAG)). Just as we can apply these methods to induce factual responses, we can apply the same technique to guide the model toward equitable and inclusive outcomes.

For example, consider an online news outlet looking to use an LLM to review article content for grammar and readability. The model does an excellent job of reviewing and revising its drafts. However, during peer review, it realizes some of the language is culturally insensitive or lacks inclusivity. As discussed, qualitative evaluation and human oversight are critical to ensuring that model output aligns with human judgment. Notwithstanding, the writing team can guide the model toward alignment with company values using a set of general guidelines for inclusive and debiased language. For example, it could ground the model with excerpts from its internal policy documents or content from its unconscious bias training guides.

Employing methodologies such as RLHF and DPO, alongside grounding techniques, ensures that LLMs generate content that is not only factual but also ethically aligned, demonstrating the potential of generative AI to adhere to high standards of truthfulness and inclusivity. Although we cannot underestimate or deemphasize the importance of human judgment in shaping model outputs, we can apply practical supplemental methods such as grounding to reduce the likelihood of harmful or biased model outputs.

In the next section, we'll explore the risks and ethical dilemmas posed by attempts to circumvent the constraints we have just discussed, highlighting the ongoing challenge of balancing the rapid adoption of generative LLMs with appropriate safeguards against misuse.

Understanding jailbreaking and harmful behaviors

In the context of generative LLMs, the term **jailbreaking** describes techniques and strategies that intend to manipulate models to override any ethical safeguards or content restrictions, thereby enabling the generation of restricted or harmful content. Jailbreaking exploits models through sophisticated adversarial prompting that can induce unexpected or harmful responses. For example, an attacker might try to instruct an LLM to explain how to generate explicit content or express discriminatory views. Understanding this susceptibility is crucial for developers and stakeholders to safeguard applied generative AI against misuse and minimize potential harm.

These jailbreaking attacks exploit the fact that LLMs are trained to interpret and respond to instructions. Despite sophisticated efforts to defend against misuse, attackers can take advantage of the complex and expansive knowledge embedded in LLMs to find gaps in their safety precautions. In particular, models that have been trained on uncurated datasets are the most susceptible, as the universe of possible outputs that the models sample from can include harmful and toxic content. Moreover, LLMs are multilingual and can accept various encodings as input. For example, an encoding such as **base64**, which can be used to translate plain text into binary format, could be applied to obfuscate a harmful instruction. In this case, safety filters may perform inconsistently, failing to detect some languages or alternative inputs.

Despite this inherent weakness in LLMs, developers and practitioners can take several practical steps to mitigate jailbreaking risks. Remember, these cannot be exhaustive as new adversarial techniques are often uncovered:

- **Preprocessing and safety filtering**: Implement robust content filtering to detect and block unsafe semantic patterns across languages and input types. For example, a firm might apply machine learning techniques to analyze prompts for adversarial patterns and block suspicious inputs before passing them to the LLM.

- **Postprocessing and output screening**: Apply a specialized classifier or other sophisticated technique to screen LLM outputs for inappropriate content before returning them.

- **Safety-focused fine-tuning**: Provide additional safety-focused fine-tuning to the LLM to reinforce and expand its safety knowledge. Focus on known jailbreaking tactics.

- **Monitoring and iterating**: Actively monitor for jailbreaking or policy violation attempts in production, analyze them to identify gaps, and continually update defense measures to stay ahead of creative attackers.

While eliminating all possible jailbreaking attempts is infeasible, a multi-layered defense and operational best practices can significantly mitigate the risk.

In the next section, we will apply a real-time defense mechanism for jailbreaking, all while reducing the likelihood of biased and harmful output.

Practice project: Minimizing harmful behaviors with filtering

For this project, we will use response filtering to try to minimize misuse and curb unwanted LLM output. Again, we'll consider our hypothetical business, StyleSprint. After successfully using an LLM to generate product descriptions and fine-tuning it to answer FAQs, StyleSprint now wants to attempt to use a general-purpose LLM (without fine-tuning) to refine its website search. However, giving its customers direct access to the LLM poses the risk of misuse. Bad actors may attempt to use the LLM search to produce harmful content with the intention of harming StyleSprint's reputation. To prevent this behavior, we can revisit our RAG implementation from *Chapter 7*, applying a filter that evaluates whether queries deviate from the appropriate use.

Reusing our previous implementation from the last chapter (found in the GitHub repository: `https://github.com/PacktPublishing/Generative-AI-Foundations-in-Python`), which applied RAG to answer specific product-related questions, we can evaluate how the model would respond to questions outside the desired scope. Recall that RAG is simply a vector search engine combined with an LLM to produce coherent and more precise responses, contextualized by a specific data source. We will directly reuse that implementation and the same product data for simplicity, but this time, we'll input a completely unrelated query instead of asking about products:

```
# random query
response = query_engine.query("describe a giraffe")
print(response)
=> A giraffe is a tall mammal with a long neck, distinctive spotted
coat, and long legs. They are known for their unique appearance and
are the tallest land animals in the world.
```

As we can see, the model did not attempt to constrain its answer to the contents of the search index. It returned an answer based on its vast training. This is precisely the behavior we want to avoid. Imagine that a bad actor induced the model to produce explicit content or some other unwanted output. Moreover, consider a sophisticated attacker that could induce the model to leak training data or expose sensitive information accidentally memorized during training procedures (Carlini et al., 2018; Hu et al., 2022). In either case, StyleSprint could face material risk and exposure.

To prevent this, we can leverage a filter to constrain the output to provide answers relevant to a given question explicitly. The implementation is already built into the LlamaIndex RAG interface. It is a feature they call Structured Answer Filtering:

> *With structured_answer_filtering set to True, our refine module is able to filter out any input nodes that are not relevant to the question being asked. This is particularly useful for RAG-based Q&A systems that involve retrieving chunks of text from external vector store for a given user query. (LlamaIndex)*

In short, this functionality gives us fine-grained control to restrict the context we provide to the LLM for synthesis, ensuring that only the most relevant results are included. Filtering out irrelevant content before synthesizing responses ensures that only information related to the user's question is used. This approach helps avoid answers that are off-topic or outside the intended subject matter. We can quickly reimplement our RAG approach, applying minor changes that enable the feature.

> **Note**
> This functionality is most reliable when using an LLM that can support function calling.

Let's see how this functionality can be implemented.

```
from llama_index.core import get_response_synthesizer
from llama_index.core.retrievers import VectorIndexRetriever
from llama_index.core.query_engine import RetrieverQueryEngine

# Configure retriever
retriever = VectorIndexRetriever(index=index,similarity_top_k=1)

# Configure response synthesizer
response_synthesizer = get_response_synthesizer(
    structured_answer_filtering=True,
    response_mode="refine"
)

# Assemble query engine
safe_query_engine = RetrieverQueryEngine(
    retriever=retriever,
    response_synthesizer=response_synthesizer
)

# Execute query and evaluate response
print(safe_query_engine.query("describe a summer dress with price"))
# => A lightweight summer dress with a vibrant floral print, perfect
for sunny days, priced at 59.99.
```

```
print(safe_query_engine.query("describe a horse"))
# => Empty Response
```

Using this approach, the model returns a response to the standard question but no response to the irrelevant question. In fact, we can take this further and compound this filtering with additional instructions in the prompt template. For example, if we revise `response_synthesizer`, we can promote a stricter response from the LLM:

```
QA_PROMPT_TMPL = (
    "Context information is below.\n"
    "--------------------\n"
    "{context_str}\n"
    "--------------------\n"
    "Given only the context information and no prior knowledge, "
    "answer the query.\n"
    "Query: {query_str}\n"
    "Answer: "
    "Otherwise, state: I cannot answer."
)
STRICT_QA_PROMPT = PromptTemplate(
    QA_PROMPT_TMPL, prompt_type=PromptType.QUESTION_ANSWER
)

# Configure response synthesizer
response_synthesizer = get_response_synthesizer(
    structured_answer_filtering=True,
    response_mode="refine",
    text_qa_template=STRICT_QA_PROMPT
)
```

This time, the model responded explicitly, `I cannot answer`. Using a prompt template, StyleSprint could return a message it deems appropriate in response to inputs unrelated to the search index and, as a side effect, ignore queries that do not adhere to its policies. Although not entirely a perfect solution, combining RAG with more strict answer filtering can help deter or defend against harmful instructions or adversarial prompting. Additionally, as explored in *Chapter 7*, we can apply RAG-specific evaluation techniques such as RAGAS to measure factual consistency and answer relevancy.

Summary

In this section, we recognized the increasing prominence of generative AI and explored the ethical considerations that should steer its progress. We outlined key concepts such as transparency, fairness, accountability, respect for privacy, informed consent, security, and inclusivity, which are essential to the responsible development and use of these technologies.

We reviewed strategies to attempt to counter these biases, including human-aligned training techniques and practical application-level measures against susceptibilities such as jailbreaking. In sum, we explored a multidimensional and human-centered approach to generative AI adoption.

Having completed our foundational exploration of generative AI, we can now reflect on our journey. We began by laying the groundwork, examining foundational generative architectures such as generative adversarial networks (GANs), diffusion models, and transformers.

Chapters 2 and *3* guided us through the evolution of language models, with a particular focus on autoregressive transformers. We explored how these models have significantly advanced the capabilities of generative AI, pushing the boundaries of machine understanding and the generation of human-like language.

Chapter 4 provided us with practical experience in production-ready environments. In *Chapter 5*, we explored the fine-tuning of LLMs for specific tasks, a technique that enhances their performance and adaptability to specific applications. *Chapter 6* focused on the concept of domain adaptation, demonstrating how tailoring AI models to understand domain-specific nuances can greatly improve their utility in specialized fields such as finance, law, and healthcare.

Chapters 7 and *8* centered on prompt engineering and constrained generation, addressing techniques to ensure that AI-generated content remains trustworthy and aligned with ethical guidelines.

This book has aimed to provide a solid foundation in generative AI, preparing professionals across disciplines and sectors with the necessary theoretical knowledge and practical skills to effectively engage with this transformative technology. The potential of generative AI is significant, and with our deeper understanding of its technologies, coupled with a thoughtful approach to ethical and societal considerations, we are ready to responsibly leverage its advantages.

References

This reference section serves as a repository of sources referenced within this book; you can explore these resources to further enhance your understanding and knowledge of the subject matter:

- Sun, L., Huang, Y., Wang, H., Wu, S., Zhang, Q., Gao, C., Huang, Y., Lyu, W., Zhang, Y., Li, X., Liu, Z., Liu, Y., Wang, Y., Zhang, Z., Kailkhura, B., Xiong, C., Xiao, C., Li, C., Xing, E., . . . Zhao, Y. (2024). *TrustLLM: Trustworthiness in Large Language Models. ArXiv*. /abs/2401.05561

- Carlini, N., Liu, C., Erlingsson, Ú., Kos, J., & Song, D. (2018). *The secret sharer: Evaluating and testing unintended memorization in neural networks*. In arXiv [cs.LG]. `http://arxiv.org/abs/1802.08232`

- Hu, H., Salcic, Z., Sun, L., Dobbie, G., Yu, P. S., & Zhang, X. (2022). *Membership inference attacks on machine learning: A survey. ACM Computing Surveys*, 54(11s), 1–37. `https://doi.org/10.1145/3523273`

- LlamaIndex. (n.d.). *Response synthesizers. In LlamaIndex Documentation (stable version)*. Retrieved March 12, 2024. `https://docs.llamaindex.ai/en/stable/module_guides/querying/response_synthesizers/root.html`

Index

packtpub.com

Subscribe to our online digital library for full access to over 7,000 books and videos, as well as industry leading tools to help you plan your personal development and advance your career. For more information, please visit our website.

Why subscribe?

- Spend less time learning and more time coding with practical eBooks and Videos from over 4,000 industry professionals
- Improve your learning with Skill Plans built especially for you
- Get a free eBook or video every month
- Fully searchable for easy access to vital information
- Copy and paste, print, and bookmark content

Did you know that Packt offers eBook versions of every book published, with PDF and ePub files available? You can upgrade to the eBook version at packtpub.com and as a print book customer, you are entitled to a discount on the eBook copy. Get in touch with us at customercare@packtpub.com for more details.

At www.packtpub.com, you can also read a collection of free technical articles, sign up for a range of free newsletters, and receive exclusive discounts and offers on Packt books and eBooks.

Other Books You May Enjoy

If you enjoyed this book, you may be interested in these other books by Packt:

Mastering NLP from Foundations to LLMs

Lior Gazit, Meysam Ghaffari

ISBN: 978-1-80461-918-6

- Master the mathematical foundations of machine learning and NLP Implement advanced techniques for preprocessing text data and analysis Design ML-NLP systems in Python

- Model and classify text using traditional machine learning and deep learning methods

- Understand the theory and design of LLMs and their implementation for various applications in AI

- Explore NLP insights, trends, and expert opinions on its future direction and potential

OpenAI API Cookbook

Henry Habib

ISBN: 978-1-80512-135-0

- Grasp the fundamentals of the OpenAI API
- Navigate the API's capabilities and limitations of the API
- Set up the OpenAI API with step-by-step instructions, from obtaining your API key to making your first call
- Explore advanced features such as system messages, fine-tuning, and the effects of different parameters
- Integrate the OpenAI API into existing applications and workflows to enhance their functionality with AI
- Design and build applications that fully harness the power of ChatGPT

Packt is searching for authors like you

If you're interested in becoming an author for Packt, please visit `authors.packtpub.com` and apply today. We have worked with thousands of developers and tech professionals, just like you, to help them share their insight with the global tech community. You can make a general application, apply for a specific hot topic that we are recruiting an author for, or submit your own idea.

Share Your Thoughts

Now you've finished *Generative AI Foundations in Python*, we'd love to hear your thoughts! Scan the QR code below to go straight to the Amazon review page for this book and share your feedback or leave a review on the site that you purchased it from.

https://packt.link/r/1-835-46082-8

Your review is important to us and the tech community and will help us make sure we're delivering excellent quality content.

Download a free PDF copy of this book

Thanks for purchasing this book!

Do you like to read on the go but are unable to carry your print books everywhere?

Is your eBook purchase not compatible with the device of your choice?

Don't worry, now with every Packt book you get a DRM-free PDF version of that book at no cost.

Read anywhere, any place, on any device. Search, copy, and paste code from your favorite technical books directly into your application.

The perks don't stop there, you can get exclusive access to discounts, newsletters, and great free content in your inbox daily

Follow these simple steps to get the benefits:

1. Scan the QR code or visit the link below

https://packt.link/free-ebook/9781835460825

2. Submit your proof of purchase
3. That's it! We'll send your free PDF and other benefits to your email directly

www.ingramcontent.com/pod-product-compliance
Lightning Source LLC
Chambersburg PA
CBHW080529060326
40690CB00022B/5075